Hilkje C. Hänel, Fabian Schuppert (eds.)
Understanding Social Struggles

Philosophy – Enlightenment – Critique | 13

Editorial

The book series **Philosophy – Enlightenment – Critique** aims to promote philosophical thinking dedicated to a future worth living for everyone in times of global crisis. Climate change, political and religious authoritarianism, and growing social inequalities – the manifold and interconnected issues of our time require a return to the power of reason. In the spirit of a new enlightenment, the series initiates a dialog between different philosophical schools and traditions that critically scrutinize the past and present and explore the ramifications of sustainable alternatives. This requires a re-examination, re-interpretation and revision of the philosophical canon. The series also reveals the emancipatory potential inherent in the interplay between philosophy and other disciplines such as technology or aesthetics.

Hilkje C. Hänel (Dr. phil.) teaches and researches at the University of Potsdam in political and social epistemology, feminist philosophy, political theory and applied ethics with a focus on migration and disability. She is co-PI of the International Research Network on the Relation of Theories of Epistemic Injustice and Recognition Theory, funded by the DFG.
Fabian Schuppert (Prof.) teaches and researches political theory and philosophy at Universität Potsdam. He is co-PI of the International Research Network on The Relation of Theories of Epistemic Injustice and Recognition Theory, funded by the DFG.

Hilkje C. Hänel, Fabian Schuppert (eds.)

Understanding Social Struggles

Relating Recognition Theories and Epistemic Injustice

[transcript]

The book was written and edited as part of the DFG project "Investigating the Relationship between Theories of Recognition and Theories of Epistemic Injustice" (455582105) and was completed and published with the help of this funding.

Bibliographic information published by the Deutsche Nationalbibliothek
The Deutsche Nationalbibliothek lists this publication in the Deutsche Nationalbibliografie; detailed bibliographic data are available in the Internet at https://dnb.dnb.de

transcript Verlag | Hermannstraße 26 | D-33602 Bielefeld | live@transcript-verlag.de

Cover design: Maria Arndt
Cover illustration: svetolk / AdobeStock
Printing: Elanders Waiblingen GmbH, Waiblingen
https://doi.org/10.14361/9783839400050
Print-ISBN: 978-3-8376-5217-8 | PDF-ISBN: 978-3-8394-0005-0
ISSN of series: 2941-8151 | eISSN of series: 2941-816X

Printed on permanent acid-free text paper.

Contents

Musings

Understanding Social Struggles

On the Companionship of Theories of Epistemic Injustice and Recognition Theory[1]

Hilkje C. Hänel & Fabian Schuppert

In recent years, there has been growing scholarly interest in the relationship between epistemic injustice and the concept of recognition. Although emerging from distinct traditions—analytic social epistemology on the one hand, and critical theory on the other—these frameworks converge on a central philosophical concern: the normative and political dimensions of subjectivity within social relations. Researchers have begun exploring how these two concepts intersect, with important contributions from scholars such as Congdon (2017), Giladi (2018), and Hänel (2020), among others. While much of this work has been productive in highlighting the role of recognition in addressing epistemic wrongs, the precise nature of the causal or constitutive relation between epistemic injustice and recognition theory remains underexplored. In the introduction, we aim to review the current literature, before briefly

1 All contributions to this book are the outcome of seminars, discussions, and workshops that took place between 2021 and 2024 and that were organized by the DFG-funded research network (project 455582105) on the relation between theories of epistemic injustice and recognition theory. We are grateful to all members of the research network who contributed in many different ways to the project: Jacob Blumenfeld, Christine Bratu, Robin Celikates, Matthew Congdon, Emmalon Davis, Franziska Dübgen, Kristina Lepold, Gonçalo Marcelo, José Medina, Karen Ng, Kerstin Reibold, Breno Santos, Briana Toole, and Rosie Worsdale, as well as the many contributors to our workshops and conferences: Daniel Loick, Candice Delmas, Nathan Rochelle DuFord, Rocío Zambrana, Adrian Wilding, Linda Alcoff, Ezgi Sertler, Paul Taylor, Nadja El Kassar, Thomas Khurana, Paul Giladi, Eraldo Souza dos Santos, Eric-John Russell, Lindsay Crawford, and Sergio Gallegos (in no particular order). We are also deeply grateful to Nikolaus Hoffmann, who has helped enormously with proofreading and preparing all contributions for publication.

discussing some of the key points that have structured our discussion on this topic over the last five years and carving out a clearer understanding of the relationship of failures of recognition and epistemic injustice by critically engaging with existing theories. We finish with an overview of the chapters in this book.

1. Brief Literature Review

A review of the literature reveals two key points of concern. First, while theorists exhibit a sophisticated grasp of epistemic injustices—those wrongs related to the denial of knowledge, credibility, or authority—they tend to approach recognition in a more intuitive or undeveloped manner. Second, when recognition theory is addressed in greater detail, much of the focus centers around a Honnethian conception of recognition, which may limit the scope of inquiry into other potentially valuable perspectives. This paper argues that a broader engagement with diverse theoretical frameworks of recognition could shed new light on the constitutive role recognition plays in epistemic injustice.

We propose that there are at least three additional, fruitful understandings of recognition that deserve closer examination in relation to epistemic injustice. First, we explore a Marxist-Indigenous understanding of recognition, drawing on the work of Glen Coulthard, which emphasizes the intersection of colonialism, indigeneity, and epistemic silencing. Second, we suggest an Hegelian-Republican perspective, influenced by philosophers like Robert Brandom and John Pippin, which highlights the role of social practices and political recognition in shaping knowledge. Finally, we introduce a feminist understanding grounded in Hegelian thought, which challenges gendered epistemic injustices and the ways in which recognition structures knowledge within patriarchal frameworks. By exploring these distinct theoretical lenses, we aim to clarify the ways in which recognition can both enable and constitute epistemic justice, offering new avenues for understanding and addressing epistemic injustice in contemporary society.

1.1 Causal and Constitutive Relations

The relationship between epistemic injustice and recognition failures is a central concern in contemporary philosophical debates. While epistemic injustice, as conceptualized by Fricker (2007), concerns the wrongful treatment of individuals in their capacity as knowers, recognition theory, as developed in the works of Honneth (1995) and others, highlights the role of intersubjective acknowledgment in the constitution of individual identity and agency. The intersection of these domains raises important questions about the nature of their connection: Are epistemic injustices primarily effects of antecedent recognition failures, or do they themselves function as acts of misrecognition? Is the relationship between these phenomena best understood causally, constitutively, or in some hybrid manner?

In assessing these questions, five guiding inquiries structure the current discourse:

(1) Is there a causal relation between epistemic injustice and failures of recognition such that epistemic injustices result from a prior history of recognition failures?
(2) Is there a causal relation between epistemic injustice and failures of recognition such that epistemic injustices render speakers vulnerable to future acts of recognition failures?
(3) Is there a causal relation between epistemic injustice and failures of recognition such that epistemic injustices (re-)produce ongoing misunderstandings of the social identity of marginalized and oppressed groups?
(4) Is there a constitutive relation between epistemic injustice and failures of recognition such that epistemic injustices constitute recognition failures (or some forms of recognition failures)?
(5) Is there a constitutive relation between epistemic injustice and failures of recognition such that recognition failures (or some recognition failures) constitute epistemic injustices?

Several scholars have proposed causal explanations linking epistemic injustice and recognition failures. Hänel (2020) argues for the first two causal relations, suggesting that epistemic injustice can be traced to historical patterns of misrecognition and that such injustices perpetuate further recognition failures. This account aligns with broader structural critiques of epistemic marginalization, wherein entrenched social hierarchies sustain cycles of epistemic exclusion. Similarly, McConkey (2004) advances the third causal relation, contending that epistemic injustice fosters distorted or inadequate representations of marginalized identities, thereby reinforcing societal misrecognition and misunderstanding.

Beyond causal explanations, other scholars maintain that the relationship between epistemic injustice and recognition failures is constitutive rather than sequential. Congdon (2018) articulates a position grounded in the moral dimensions of epistemic agency, arguing that being a knower entails a normative status that is both epistemic and ethical. As Congdon explains:

> "It is epistemic insofar as the label 'knower' indicates the roles one may legitimately assume within practices of justification and warrant, and ethical, in the sense that being a knower implicates one within interpersonal relations of answerability that invoke notions of justice and injustice, flourishing and degradation, virtue and vice, rightful treatment and moral injury" (2018, 2).

If epistemic agency is embedded within a broader moral framework, then being denied recognition as a knower is inherently a recognition failure. Such denial undermines one's normative standing within the epistemic community, restricting access to practices of justification and eroding the social preconditions for epistemic flourishing.

Similarly, Giladi (2017) emphasizes the role of epistemic respect in fostering individual self-confidence as rational enquirers. According to this view, recognition is not merely an external validation but a necessary condition for epistemic self-trust and participation. Failure to extend such recognition, then, is not simply an epistemic injustice but a fundamental act of misrecognition that shapes one's epistemic identity and agency.

These perspectives suggest that recognition theory is not merely related to epistemic concerns but is deeply embedded within them. The denial of recognition, whether as an act of misrecognition or as an effect of epistemic injustice, threatens one's standing as a legitimate source of knowledge and reason. The next section will explore how these insights inform contemporary approaches to epistemic justice, recognition, and the ethics of knowledge exchange in diverse social and political contexts.

1.2 Intuitive Notions of Recognition

Discussions of epistemic injustice, on the other hand, often rely on an intuitive understanding of recognition. However, these discussions sometimes diverge from or even obscure traditional recognition theory's deeper normative commitments. This section examines how intuitive notions of recognition function in epistemic injustice literature and whether these usages align with or distort recognition theory's epistemic implications.

Some scholars argue that epistemic injustice stems from recognition failures. However, Mikkola (2022) challenges this assumption by contending that ignorance and recognition can sometimes function cooperatively. Drawing on Mills' racial contract, Mikkola suggests that the exclusion of people of color from full citizenship does not merely involve ignorance but also an implicit affirmation of normatively relevant properties of those excluded. In this way, recognition is present, but it serves an exclusionary function rather than one of epistemic validation.

Further cases of intuitive recognition arise in the work of Jackson (2022) and Freeman & Stewart (2022), who analyze epistemic injustices related to recognition failures in contexts of sexual violence and medical treatment. Jackson (2022) explores how men who have experienced sexual violence often face epistemic injustice due to societal reluctance to recognize them as victims. Freeman & Stewart (2022) extend this analysis to transgender and gender non-binary patients in medical contexts, arguing that misrecognition—such as failing to recognize a patient's gender identity—results in epistemic harms. While these discussions employ recognition in an intuitive sense, they sometimes lack a systematic engagement with recognition theory's deeper commitments.

Traditional recognition theory, particularly in the works of Honneth, frames recognition not merely as acknowledgment but as a necessary condition for subject-formation. That is, recognition is not only about being seen but about being conferred a normative status that underlies self-relation and agency. Misrecognition, therefore, does more than cause epistemic harm—it disrupts the very conditions under which one can develop a coherent sense of self.

This insight suggests that while epistemic injustice literature can illuminate aspects of recognition failures, it risks distorting recognition theory's epistemic dimensions by treating recognition too loosely. A more robust engagement with recognition theory would enhance the analysis of epistemic injustices by situating them within a broader framework of intersubjective dependence and normative subject-formation.

1.3 Honnethian Understandings of Recognition

Several scholars have drawn on Honneth's framework to interpret epistemic injustice as a form of misrecognition. Giladi (2017) argues that testimonial and hermeneutical injustices function as failures of recognition. Testimonial injustice denies marginalized speakers credibility, thereby undermining their status as rational agents, while hermeneutical injustice alienates individuals from collective epistemic resources, preventing them from fully participating in meaning-making processes. Giladi situates these injustices within Honneth's broader conception of misrecognition as a social pathology that obstructs self-realization.

Congdon (2017, 2018) similarly argues that epistemic injustice constitutes a form of misrecognition by eroding an individual's self-relation. Because recognition is essential for self-confidence, self-respect, and self-esteem, epistemic injustice disrupts these dimensions of practical self-understanding. Meanwhile, Petherbridge (2022) contends that epistemic injustice highlights a distinct epistemic dimension of recognition theory that precedes Honneth's three-tier model of love, rights, and solidarity.

While these interpretations enrich discussions on epistemic injustice, they also raise questions about whether Honneth's framework sufficiently captures the full complexity of epistemic harms. Other recognition theorists may offer alternative insights that better account for the nuances of epistemic injustice.

2. The Discussion

Many of the contributions in this edited collection start from the insight that epistemic injustice can be fruitfully understood as a form of misrecognition or that a theory of epistemic justice is best situated within a broader recognitional framework; yet, these insights are developed into significantly different theories of the role that theories of epistemic injustice and recognition theory play in our philosophical attempts to understand social struggles and movements. In doing so, the contributions seek to bridge gaps between moral psychology, epistemology, and social theory, offering accounts of how subjects are wronged in their capacities as knowers, agents, and members of social communities and in how they resist such wrongs. Before we provide a brief summary of the chapters, let us point out some of the key themes that have emerged during our discussions on the topic.

2.1 Epistemic Injustice: Beyond the Cognitive Harm

Miranda Fricker's work has been seminal in identifying two central forms of epistemic injustice: testimonial injustice and hermeneutical injustice. Testimonial injustice occurs when a speaker's credibility is unjustly deflated due to prejudice, typically based on race, gender, or class. Hermeneutical injustice, by contrast, arises when a marginalized group lacks access to the shared conceptual resources needed to make sense of their experiences. Both are epistemic harms, yet Fricker insists they also bear moral weight: they wrong individuals in their capacities as knowers.

However, critiques of Fricker's account point to certain limitations. Scholars such as Kristie Dotson (2011) and Gaile Pohlhaus Jr. (2017) argue that Fricker under-theorizes the structural and affective dimensions of epistemic injustice, focusing too narrowly on individual virtuous responses. They call for a broader account that incorporates historical power structures and group-based epistemic agency. This shift opens the door to thinking about epistemic injustice not just as a deficit in knowledge practices but as a deeper form of social and moral misrecognition.

2.2 Recognition Theory: Subjectivity and Social Justice

Axel Honneth's recognition theory, building on Hegelian and post-Hegelian traditions, posits that individual identity is formed and sustained through intersubjective recognition. Honneth distinguishes three spheres of recognition: love (emotional support), rights (legal respect), and solidarity (social esteem). Failures in recognition—be they emotional neglect, denial of rights, or social devaluation—amount to moral injuries that impede the development of self-confidence, self-respect, and self-esteem.

Recognition is not merely a matter of individual perception but is embedded within social institutions and cultural norms. Misrecognition thus entails more than subjective insult; it is a distortion of social relations that undermines personhood. From this perspective, the moral and psychological harm inflicted by misrecognition bears a strong resemblance to the epistemic harm described by Fricker and her interlocutors.

2.3 Epistemic Injustice as Misrecognition

The concept of recognition allows us to reframe epistemic injustice as a specific modality of misrecognition. When someone is treated as lacking epistemic credibility due to their social identity, they are not merely being underestimated intellectually; they are being denied recognition as a rational and trustworthy agent. Testimonial injustice, then, is not only an epistemic failure but also a failure of intersubjective respect.

Similarly, hermeneutical injustice reflects a lack of cultural recognition. When collective interpretive resources fail to include the experiences of marginalized groups, those groups are rendered unintelligible in the public sphere. This form of epistemic marginalization parallels the cultural misrecognition theorized by Charles Taylor and Honneth, wherein entire forms of life are excluded from social visibility and esteem.

This integrated account reveals that epistemic injustice undermines the very conditions of recognitional justice. To be denied credibility is to be denied moral respect; to lack hermeneutical resources is to be excluded from cultural belonging. Thus, epistemic injustice can be viewed as a double injury: it deprives individuals of both epistemic and moral standing.

2.4 Epistemic Struggles and Social Movements

Epistemic injustice and misrecognition are not merely philosophical abstractions—they are lived experiences that often animate social struggles. Movements for racial justice, gender equality, Indigenous rights, and disability justice are all, in part, struggles for epistemic recognition. Activists contest dominant narratives, demand credibility for marginalized knowers, and create new hermeneutical resources that render their experiences intelligible and politically salient.

For example, the feminist slogan "the personal is political" can be read as a collective effort to overcome hermeneutical injustice by naming and theorizing previously silenced experiences of oppression. Importantly, not merely by adding new or revised concepts to the dominant hermeneutical resources—as Fricker suggests in her account—but by shifting the entire epistemic system such that the statement becomes intelligible; thus, working in line with what Dotson analyzes as third-degree changes, where it is not sufficient to change our conceptual resources within a given epistemic system but where we have to change the system as such. Similarly, the Black Lives Matter movement not only calls attention to racialized violence but also insists on the credibility and authority of Black voices in public discourse. These movements are epistemic interventions as much as they are political ones.

While recognition theory can help us understand these struggles as demands for visibility, respect, and inclusion in the interpretive and discursive frameworks of society, the task at hand is to dig deeper and to understand the call for changing the epistemic system entirely also for a call to change the entire recognition system. These struggles are not merely about legal rights or material redistribution, but about the transformation of the conditions under which individuals and groups are acknowledged as legitimate sources of knowledge and meaning.

Moreover, as José Medina and María Lugones have emphasized, epistemic resistance is a crucial element of (democratic) struggles. Marginalized communities not only suffer epistemic injustice but also develop counter-knowledges and resistant epistemologies—what Medina calls "epistemic counterpoints"—that challenge the dominant epistemic order. These counterpoints often emerge through collective praxis, storytelling, and grassroots organizing, and they serve to expand the epistemic imagination of society as a whole and to open the possibility for new systems.

Indigenous knowledge movements, for instance, not only seek land rights or political sovereignty but also advocate for the recognition of alternative epistemologies rooted in relationality, oral tradition, and ecological stewardship. The denial of such knowledge systems constitutes a deep form of epistemic colonialism and misrecognition. A recognitional approach helps us see that justice for Indigenous communities must include epistemic justice: the validation and inclusion of their ways of knowing within the broader epistemic community.

2.5 Normative Implications

Understanding epistemic injustice as misrecognition has important normative consequences. It shifts the locus of moral concern from individual cognitive failings to the institutional and cultural frameworks that sustain exclusion. It also demands a more comprehensive account of justice—one that encompasses not only distributive and legal dimensions, but also recognitive and epistemic ones.

In this view, epistemic justice entails the restructuring of educational systems, media representations, legal discourses, and other public institutions to affirm the epistemic agency of all individuals and communities. It also calls for cultivating affective and attitudinal dispositions conducive to recognition—such as openness, humility, and solidarity—within epistemic practices.

3. Expanding Recognition Theory: Alternative Approaches

The general idea of this book then is that theories of epistemic injustice and recognition can and should be integrated into a unified normative framework; yet, we have to proceed with caution as not to oversee the way in which our current dominant epistemic system and recognition system makes certain things unintelligible and posits certain social groups outside of being bearers of recognition. The framework we have in mind allows us to understand epistemic harms not only as failures of knowledge but as moral injuries rooted in misrecognition. By situating epistemic justice within the broader project of recognitive justice, we gain a richer, more socially grounded account of what it means to be wronged as a knower—and of what it would mean to be justly recognized as one.

In a world marked by deep epistemic inequality, this integration is not only conceptually fruitful but morally urgent. As such, philosophers must continue to explore the intersection of epistemology and recognition, illuminating the pathways by which knowledge, identity, and justice are inextricably linked. With this book, we hope to open the room for these discussions. Before we let the contributions of this book speak for themselves, we want to finish this introduction by raising some problems that might stand in the way of an emancipatory analysis of social struggles via the tools of recognition theory and theories of epistemic injustice. In doing so, we aim to examine the relationship between epistemic injustice and recognition theory, critically engaging with Marxist-indigenous, Hegelian-republican, and feminist perspectives. Its aim is to go beyond the current debate, following recent critiques both of Honnethian recognition theory and Frickerian epistemic injustice theory. For example, while Honneth's theory emphasizes recognition as essential for self-realization, alternative approaches reveal its complicity in structural oppression. Instead, Coulthard critiques liberal recognition for reinforcing colonial and capitalist domination. The Hegelian-republican tradition highlights hierarchical misrecognition and its role in maintaining inequality. And, feminist theorists foreground vulnerability and differential epistemic visibility. By synthesizing these perspectives, we reframe recognition as both a mechanism of epistemic injustice and a potential site of resistance, linking it to broader struggles of social and epistemic transformation. Yet, we do not intent to give a conclusive argument for a specific relationship between epistemic injustice and failures of recognition; rather, we aim to raise key questions that any engagement with these theories should take seriously.

3.1 Beyond Honneth's Theory of Recognition

Beyond Honneth's framework, alternative recognition theories offer valuable insights into epistemic injustice. Marxist-indigenous critiques highlight the socioeconomic dimensions of recognition, arguing that recognition must be understood within the broader context of material power relations (Coulthard 2014). Hegelian-republican interpretations emphasize recognition as integral to social freedom and equality, challenging hierarchical forms of recognition (Brandom 2019; Forst 2017). Feminist perspectives introduce the notion of vulnerability as central to recognition, arguing that recognition operates within structures of power and differential exposure to harm (Oliver 2001, 2015; Petherbridge 2016).

These perspectives collectively underscore two critical insights: (1) recognition and vulnerability are distributed unequally across social structures, and (2) the struggle for recognition is not necessarily best understood as a struggle for recognition from those in power. Engaging these alternative models expands the scope of recognition theory and provides a richer framework for analyzing epistemic injustice, particularly in contexts of oppression and resistance. Future research should further integrate these diverse perspectives to develop a more comprehensive understanding of recognition and epistemic justice.

The relationship between recognition and epistemic injustice becomes even more complex when examined through alternative theoretical lenses. While traditional recognition theory, particularly as formulated by Honneth, has focused on intersubjective acknowledgment as essential for individual self-realization, other critical perspectives emphasize the broader structural, political, and material conditions that shape recognition practices. This article's aim was to examines the intersection of epistemic injustice and recognition theory, exploring how alternative frameworks—Marxist-indigenous, Hegelian-republican, and feminist perspectives—reconfigure our understanding of recognition as both a site of oppression and a tool of resistance.

While traditional recognition theory, particularly as developed by Honneth, emphasizes intersubjective acknowledgment as essential for self-realization, critical perspectives highlight the structural, political, and material conditions that shape recognition practices and their epistemic consequences.

First, the Marxist-indigenous critique, drawing on Coulthard (2014), challenges the liberal model of recognition for its failure to account for the dual structure of colonialism, encompassing both ideological and material dimensions. Coulthard argues that recognition functions as a mechanism of power, wherein the settler-state co-opts indigenous struggles, reinforcing colonial and capitalist relations rather than dismantling them.

Second, the Hegelian-republican approach, as developed by Brandom (2019), Pippin (2006), and Forst (2017), critiques hierarchical recognition models that reinforce epistemic and material inequalities. This perspective highlights the ways in which over-recognition of dominant groups entrenches power imbalances while misrecognition of the oppressed can function as an epistemic and political counterforce, generating subaltern spaces of resistance and knowledge production.

Third, feminist approaches, particularly those of Oliver (2001, 2015) and Petherbridge (2016), expand recognition theory by foregrounding vulnerability as a fundamental condition of intersubjectivity. Feminist theorists argue that recognition and epistemic visibility are differentially distributed, shaping whose voices are heard and legitimized within dominant epistemic frameworks. This insight aligns with standpoint epistemology and the epistemology of ignorance, revealing how epistemic injustice is embedded in broader social structures.

By integrating these alternative perspectives, we reposition recognition theory within a more critical framework, demonstrating that epistemic injustice is not only a failure of recognition but also a consequence of structural inequities that demand transformative resistance.

3.2 Marxist-Indigenous Critique of Liberal Recognition

Glen Coulthard's *Red Skin, White Masks* (2014) presents a Marxist-indigenous critique of liberal recognition, arguing that dominant frameworks of recognition fail to adequately account for the dual structure of colonialism: both its ideological dimension (the politics of recognition) and its material foundation (socioeconomic structures). In this view, recognition is not merely a question of being seen or acknowledged by the settler-state but is deeply embedded in the reproduction of colonial and capitalist power relations. Coulthard critiques the liberal model of recognition for two primary reasons:

- It assumes that recognition can be granted by those in power, positioning oppressed groups as dependent on the benevolence of the state or dominant society. This conceptualization frames recognition as a "gift" rather than a site of struggle.
- It neglects the materialist foundations of colonial and capitalist domination, failing to recognize how economic structures perpetuate epistemic and political subjugation.

For Coulthard, indigenous struggles for recognition must therefore be grounded in materialist and anti-colonial critique. He contends that for many Indigenous peoples, capitalism is not merely a parallel system but a function of colonialism, wherein recognition operates as a mechanism of power that both enables and conceals the ongoing dispossession of Indigenous lands and epistemic agency.

3.3 Hegelian-Republican Critique of Hierarchical Recognition

The Hegelian tradition of recognition theory is most commonly associated with the master-slave dialectic, yet its implications for epistemic justice extend beyond this historical paradigm. In republican and critical-Hegelian interpretations, recognition is understood not simply as a means of self-realization but as a precondition for social freedom and political agency (Brandom 2019; Pippin 2006; Forst 2017). Within this framework, recognition is not merely something the powerful bestow upon the marginalized; rather, it is a site of struggle that can be mobilized to challenge social hierarchies.

Two key insights emerge from this critical-Hegelian perspective:

- First, certain social groups receive an *excess* of recognition, particularly in the form of social esteem and authority, reinforcing epistemic and material inequalities. Over-recognition of dominant groups can distort social relations, entrenching power imbalances and epistemic authority.
- Second, the recognition struggles of marginalized groups can function as powerful counter-narratives. Subaltern communities do not merely seek recognition from the dominant; they also create alternative spaces of epistemic agency that resist and refuse oppressive norms. This understanding

highlights recognition as a political tool that extends beyond the deficit model of "seeking acknowledgment" and instead emphasizes the transformative potential of misrecognition and epistemic resistance (McBride 2013).

3.4 Feminist Approaches: Recognition, Vulnerability, and Epistemic Injustice

Feminist theory has long engaged with recognition as a critical component of justice, particularly in the context of gender, sexuality, and epistemic vulnerability. Drawing from Hegelian theory, feminist scholars argue that self-consciousness and subjectivity are not autonomous but develop through intersubjective relationships. However, this framework is expanded by the concept of vulnerability, which feminist theorists reinterpret in ways that challenge traditional notions of agency and recognition (Ferrarese 2011; Oliver 2001, 2015; Petherbridge 2016).

Three key arguments emerge in feminist discussions of recognition:

- Vulnerability is not reducible to physical harm or coercion; rather, it encompasses broader epistemic and social structures that determine who is seen, heard, and believed.
- Recognition, particularly in the context of epistemic injustice, is deeply tied to the visibility of marginalized groups. As Oliver (2001) argues, to be recognized is to be rendered visible within a given epistemic framework. However, this visibility is often conditional and constrained by dominant power structures.
- Vulnerability is differentially distributed. Certain groups are made disproportionately vulnerable by epistemic and social inequalities, a dynamic that parallels the selective distribution of recognition. Oliver (2015) notes that "some bodies are made vulnerable for the sake of the prosperity of others, " illustrating how recognition and epistemic power are stratified within social hierarchies.

Furthermore, feminist critiques of recognition intersect with standpoint epistemology and the epistemology of ignorance, demonstrating how dominant epistemic frameworks systematically exclude, distort, or appropriate marginalized knowledge. The struggle for recognition, therefore, is not merely about gaining visibility but about contesting the epistemic norms that define who is recognized as a legitimate knower.

This section explored three alternative frameworks—Marxist-indigenous, Hegelian-republican, and feminist perspectives—to show that shifting our gaze towards other, often marginalized theories, can significantly deepen our understanding of how recognition functions in contexts of oppression, resistance, and epistemic agency.

3.5 Conclusion: Recognition, Epistemic Justice, and Social Freedom

These three alternative perspectives on recognition theory—Marxist-indigenous, Hegelian-republican, and feminist—offer a more expansive framework for understanding epistemic injustice. They challenge the notion that recognition is a simple remedy for epistemic exclusion and instead reveal its entanglement with structures of power, material conditions, and vulnerability. Two central insights emerge from this discussion:

1. Recognition and vulnerability are not evenly distributed; they are shaped by broader social structures and relations of power.
2. The struggle for recognition should not be understood merely as a plea for acknowledgment from the dominant but as an epistemic and political act that can challenge and transform oppressive systems.

By incorporating these perspectives, recognition theory becomes not just a framework for understanding epistemic injustice but also a site of critical resistance. Future research should further explore the intersections between recognition, epistemic agency, and political struggle, examining how marginalized groups actively reshape the epistemic conditions of their own recognition.

4. Summary of the chapters

José Medina's chapter starts the book's critical investigation by developing an analysis of epistemic injustice as a failure of recognition or *misrecognition*. He argues that the lack of epistemic recognition involved in epistemic injustices can only be properly identified and fought against from the epistemic vantage points of grassroot social movements of liberation. More specifically, he shows that the "insider-outsider" status of the members of oppressed groups seeking liberation through grassroot social movements allows for: (1) the emergence of critical consciousness amid pervasive alienation and ideological distortion; and (2) the development of a critique that is both immanent and transcendent and can prefigure the melioration and self-transformation of social life through alternative economies of recognition. The arguments for these two claims address two of the problems facing the diagnosis of social pathologies, namely, (1) the problem of developing critical consciousness about forms of oppression that have become pervasive and invisibilized by ideological distortions and alienation; and (2) the problem of developing *transformative immanent critique* that is neither internal meliorism nor external interventionism, that is, a critique that doesn't come (entirely) from the outside (to avoid problems of paternalism and heteronomy), but doesn't remain (entirely) within the insider's perspective and is capable of transcending that perspective.

Jacob Blumenfeld explores the intersection of class and recognition theory, examining how the class structure degrades human dignity and autonomy. For Blumenfeld, class represents a moral injury to the status of being human. Unlike identity-based claims that seek recognition, the condition of class necessitates abolition. The paradox is that even though class can be diagnosed as a form of misrecognition, its rectification cannot be accomplished via recognition. To recognize class as a moral injury is to recognize that recognition cannot resolve the misrecognition of class. Undoing the epistemic injustices and misrecognitions of class rather demands the collective work of abolishing class itself.

Hilkje Hänel bridges social epistemology and Frankfurt School Critical Theory, focusing on how critical knowledge can be achieved by marginalized and oppressed knowers despite, what she calls, ideological ignorance and what could be described as false consciousness in Critical Theory's toolbox. Hänel attempts to show that ideological ignorance is not the only way in which emancipatory knowledge is blocked; instead, marginalized and oppressed knowers might decide—for many different reasons—to block access to emancipatory knowledge; a phenomenon that is sometimes labelled *strategic ignorance*. She then continues to discuss a tension between ideological ignorance and standpoint epistemology. If ideological ignorance can be tackled well by sharing knowledge and actualizing emancipatory and critical knowledge practices, it seems that any form of strategic ignorance is problematic for denying access to this deeply needed knowledge. In other words, if critical knowledge is not simply given but results from struggle and communal practices, then failing to engage in struggle or communal practices for strategic reasons blocks access to critical knowledge. It would then follow that strategic ignorance is problematic for our knowledge systems—a conclusion that Hänel resists. Instead, she argues that strategic ignorance is a contextual phenomenon and that practices of recognition and resistance should take priority over our responsibility for ameliorating the epistemic system.

Breno Santos continues with an epistemological analysis. Building on the recent literature of the debate, he examines the connections between the talk of epistemic injustice as domination and the debate concerning the recognitive expectations we have as members of epistemic communities. To do so, Santos explores Amandine Catala's formulation of hermeneutical domination and assess if it fits a recognitive model of the moral responsibilities we have towards other epistemic agents. He then argues that epistemic domination has its roots in failures of recognition. These failures, as he'll show, affect the agent's freedom and autonomy as a rational enquirer. Thus, the way out of them will need to involve non-domination in the form of recognition and discursive control, something that Santos will argue for, before sketching a possible global version of hermeneutical domination that challenges the efficacy of local recognition and discursive control. Institutional remedies to this global type of epistemic injustice might not be able to account for the extent to which it affects our shared space of reasons with an overarching dominating power. To deal with the structural and institutional aspects of epistemic injustices and to secure non-domination and proper recognition, then, we'll have to go beyond the moderate, intra-institutional measures of deliberative democracy and propose radical measures to secure freedom and hermeneutical non-domination.

Clara Berlich's contribution investigates a specific class of cases of testimonial injustice, that is, cases, in which speaker and hearer stand in a non-trivial relation to each other – such as friends do, or lovers, or family members. The intuition is that these cases of testimonial injustice critically exhibit more, and even a distinct kind of badness than other cases thereof. To account for this intuition, it is argued that testimonial injustice in the context of intimate relationships is not only a manifestation of injustice, but also a specific form of betrayal; bringing it closer to theories of recognition than other forms of epistemic injustice.

Peimaneh Yaghoobifarah is concerned with applying both epistemic injustice theory and recognition theory in the context of schools. The contribution argues that children in general and students in the context of schools can be understood as a separate social group. And, that they, as a result of their studenthood, can experience both testimonial and hermeneutical injustices. This marks an important distinction of multiply realizable epistemic injustice; including one that is not considered by Fricker. On the one hand, students, who are marginalized because they are marked as different from the dominant culture due to, for example, racialization or religious affiliation, can be treated in epistemically unjust ways because of their respective marginalization. This is a form of epistemic injustice well researched but not yet applied to students. On the other hand, students are additionally affected by epistemic injustice qua being students. This is a form of epistemic injustice that goes beyond the framework Fricker provides. Furthermore, the contribution shows how epistemic harms that students suffer from as a result of their marginalization or their student status can adequately be regarded as a form of misrecognition.

The second part of the book is reserved for musings; explorations of topics that have come up during our many discussions on the topic of social struggles, recognition theory and theories of epistemic injustice. These explorations aim to shed light on important topics, while also leaving room for continued discussion and development.

Karen Ng starts the musings with the attempt to develop a brief proposal how the concept of recognition can provide a helpful framework for an ethical understanding of relations within living nature; while acknowledging that there is something of an uphill battle to employing recognition as a viable concept beyond the scope of human relationships. Ng uses the term *ethical* here in a broad sense recalling Hegel's conception of ethical life or *Sittlichkeit*. Ethical relationships draw on character, dispositions, and the cultivation of habituated interactions, which gain their meaning from species-specific modes of living and can be judged as good or bad for some aim. For self-conscious creatures, ethical life and relationships can give rise to expectations or obligations, but the latter are not necessary for something to have an ethical character. Ng's suggestion will be that we can understand the relations between living beings as ethical in this broad sense, and that recognition is an appropriate concept for capturing, at least in part, the ethical character of these relationships. There are two key aspects to expanding the concept of recognition that she will pursue here. First, it is important to reestablish the essential connection between recognition and the concepts of life and love, which is most clearly articulated in Hegel's early writings. Recognition, paradigmatically expressed in love and friendship, crucially involves the apprehension of life and living form. Second, she will consider the possibility of proto-recognitive activities and processes in non-human life-forms, as well as the possibility of recognition and proto-recognition across life-forms, both human and non-human. In exploring these possibilities, what will hopefully emerge is the importance of the concept of recognition for understanding non-instrumental and ethical relationships among and between living beings, expanding the concept beyond its current, human-centric application.

Fabian Schuppert continues with an investigation of AI in the context of epistemic misrecognition. According to proponents of LLM, these technologies may seem neutral, even democratizing, as they supposedly promise to remove human bias and ensure a "level playing field." However, a closer examination reveals that LLM systems systematically reproduce—and even amplify—epistemic injustices and stereotypes about what counts as relevant knowledge, good writing, and rational argumentation. In addition, particular fields of inquiry are presented in very particular ways, meaning that existing stereotypes about what particular forms of knowledge can do and be are systematically reproduced. In doing so, LLM tools perpetuate existing power structures, making the privileged even more privileged, and foster a particularly insidious form of epistemic ignorance and epistemic domination.

References

Brandom, R.B. (2019): *A Spirit of Trust: A Reading of Hegel's Phenomenology*. HUP

Congdon, M. (2017): "What's Wrong with Epistemic Injustice? Harm, Vice, Objectification, Misrecognition." In: *The Routledge Handbook of Epistemic Injustice*.

Congdon, M. (2018): "'Knower' as an Ethical Concept: From Epistemic Agency to Mutual Recognition." *Feminist Philosophy Quarterly* 4(4).

Coulthard, G. (2014): *Red Skin, White Masks: Rejecting the Colonial Politics of Recognition*. U. of Minnesota Press.

Dotson, K. "Tracking Epistemic Violence, Tracking Practices of Silencing." *Hypatia* 26: 236–257.

Ferrarese, E. (2011): "Judith Butler's 'not particularly postmodern insight' of recognition." *Philosophy and Social Criticism* 37(7).

Freeman, L. & H. Steward (2022): "The Problem of Recognition, Erasure, and Epistemic Injustice in Medicine: Harms to Transgender and Gender Non-Binary Patients – Why We Should Be Worried." In: *Epistemic Injustice and the Philosophy of Recognition*.

Forst, R. (2017): "Noumenal Alienation: Rousseau, Kant and Marx on the Dialectics of Self-Determination." *Kantian Review* 22.4.

Giladi, P. (2017): "Epistemic Injustice: A Role for Recognition?" *Philosophy and Social Criticism* 44(2).

Giladi, P. (2022): "Epistemic Exploitation and Ideological Recognition." In: *Epistemic Injustice and the Philosophy of Recognition.*

Hänel, H.C. (2020): "Hermeneutical Injustice, (Self-)Recognition, and Academia." *Hypatia* 35(2).

Jackson, D. (2022): "Male Sexual Victimisation, Failures of Recognition, and Epistemic Injustice." In: *Epistemic Injustice and the Philosophy of Recognition.*

McBride, C. (2013): *Recognition*. Polity.

McConkey, J. (2004): "Knowledge and Acknowledgement: 'Epistemic Injustice' as a Problem of Recognition." *Politics* 24(3).

Mikkola, M. (2022): "Ideal Theory, Epistemologies of Ignorance, and (Mis)Recognition." In: *Epistemic Injustice and the Philosophy of Recognition.*

Oliver, K. (2001): *Witnessing: Beyond Recognition.* Minneapolis: U of Minnesota Press.

Oliver, K. (2015): "Witnessing, Recognition, and Response Ethics." *Philosophy & Rhetoric* 48(4).

Petherbridge, D. (2016): "What's Critical about Vulnerability? Rethinking Interdependence, Recognition, and Power." *Hypatia* 31(3).

Petherbridge, D. (2022): "A Fourth Order of Recognition? Accounting for Epistemic Injustice in Recognition Theory." In: *Epistemic Injustice and the Philosophy of Recognition.*

Pippin, R. (2006): "Recognition and Reconciliation: Actualized Agency in Hegel's Jena Phenomenology." *Hegel: New Directions* 133.

Pohlhaus, G. Jr. (2017): "Varieties of Epistemic Injustice." In: *The Routledge Handbook of Epistemic Injustice*, ed. by I.J. Kidd, J. Medine, and G. Pohlhaus. New York: Routledge.

Schuppert, F. (2015): "Non-Domination, Non-Alienation and Social Equality: Towards a Republican Understanding of Equality." *Critical Review of International Social and Political Philosophy* 18(4).

Sheth, F. (2009): *Toward a Political Philosophy of Race*. Albany: SU of New York Press.

Epistemic Injustice, Misrecognition, and Liberation Movements as Resistance Struggles for (Self-)Recognition

José Medina

Introduction: Oppression, Epistemic Marginalization, and Misrecognition

Oppressed groups typically face severe communicative and epistemic obstacles, especially when it comes to communication and knowledge about their predicaments as oppressed subjects. The visibility and audibility of members of oppressed groups when they try to protest their predicaments is often precarious and defective. Oppressed subjects are often invisible and inaudible *qua* oppressed subjects, or they are hyper-visible and hyper-audible in distorting ways.[1] Moreover, members of oppressed groups often have a hard time making their non-mainstream experiences and perspectives properly visible and audible not only to others—to out-group members—but even *to themselves*—to in-group fellow members and even to their own selves. For example, think of how difficult it is for those who are economically exploited in similar ways to develop class consciousness, to recognize themselves as *exploited* and to link their predicament to that of others as suffering similar kinds of harms that amount to an *injustice*. But think also of how difficult it was (and still is in some social contexts) for queer subjects to recognize themselves as being improperly stigmatized by heterosexist norms and expectations, to trust their own inclinations or judgements and give proper epistemic recognition to their own experiences of desire. It takes a village that doesn't yet exist, a supportive community that needs to be created, for oppressed subjects to be able to achieve vis-

1 See my discussion of the epistemic harms of invisibility and hypervisibility in Medina (In Preparation).

ibility and audibility under conditions of oppression. And this is exactly what grassroot movements of liberation do: struggles for liberation led by these social movements include struggles for recognition that can create the conditions under which certain subject positionalities can become visible and the experiences and perspectives of those subjects can receive proper recognition. And note that, as I will argue in detail throughout this essay, in these liberation struggles members of oppressed groups seek being properly recognized not only by others—by society at large and its institutions, or by members of dominant groups—*but also by themselves*, by each other and by their own emerging counter-communities of respect.

Authors like Axel Honneth (2023), Michele Moody-Adams (2022), and myself (2023) have recently called attention to the crucial role that movements of liberation play in securing social recognition for oppressed subjects and their struggles for justice, through the development of critical concepts such as "exploitation," "unpaid domestic labor," or "sexual harassment," and through liberatory slogans such as #BlackLivesMatter, #MeToo, and "We are here, we are queer, get used to it!" What I want to highlight in this essay, drawing from my recent work (Forthcoming), is that the liberatory struggles of social movements (such as the labor movement, the Women's movement, or the Queer Liberation movement) have to do not only with securing the recognition of society and its institutions, but also with making one's own *self-recognition* possible. The liberation that these movements seek require not only the mobilization of social support and recognition for their causes from out-group members, but also the mobilization of *people like them*, that is, the articulation of a public that didn't exist before and comes together through their *self-recognition* as members of a group who share similar experiences of oppression or marginalization and similar aspirations of freedom. The resistance work against deep-seated roadblocks for recognition that liberation movements engage in fits well what I have termed *epistemic resistance* (Medina 2013); and it consists in ways of making the predicaments of the oppressed visible, audible and known *not only to others, but also to themselves*. Resistance struggles for recognition involves fighting against the systematic ways in which the visibility, audibility, and knowability of the experiences of the oppressed have been blocked for all, including also (although differently and to a lesser degree) for the experiential subjects themselves: these subjects have a hard time being in touch with their own experiences of oppression and becoming able to give voice to them, not only because of the hostile communicative environments they see themselves in and the likelihood of receiving defective uptake or no uptake at all, but also and more fundamentally because they are alienated from their own experiences and subject to ideological distortions in ways that make it very difficult for them to recognize themselves as experiencing injustices. The recognition struggles undertaken by liberation movements aim at overcoming different kinds of misrecognition: being misrecognized by society and its institutions; being misrecognized by one's fellows in interpersonal interactions of all kinds; but also being misrecognized by oneself in one's own self-understanding and self-assessments.

Drawing from recent literature that brings together normative social theories of recognition and theories of epistemic injustice,[2] the rest of this paper will argue that there are deep *epistemic injustices* created by the social dynamics of misrecognition involved in systems of oppression: they include *testimonial injustices* that occur when oppressed subjects are not recognized as subjects who can be trusted when they report on their own experiences; and *hermeneutical injustices* that occur when oppressed subjects are prevented from (or undermined in their attempts of) making sense of their own experiences and predicaments in their own terms. As Miranda Fricker's own account of epistemic injustice emphasizes, when oppressed subjects are not being given the intelligibility and credibility that the interpretations and assessments of their own experiences deserve, they are unfairly treated as epistemic subjects and are, therefore, the recipients of *deficient forms of epistemic recognition* that amount to *epistemic injustices*. Other contributors[3] to the literature of epistemic injustice have emphasized that, even independently of credibility or intelligibility deficits, oppressed subjects can also suffer from *agential* epistemic injustices when their participation in meaning-making and knowledge-sharing practices is compromised because their status and agency in those practices is *not properly recognized*—here too these subjects become the recipients of deficient forms of epistemic recognition that amount to epistemic injustices.

As I will argue in the next two sections, under conditions of oppression, there are forms of non-recognition and *misrecognition* that undermine the epistemic status and agency of members of oppressed groups, and thus create epistemic injustices that make it difficult for these subjects to be properly seen and heard by others and *even by themselves*. I will develop my own view of how we should conceptualize epistemic injustices as forms of misrecognition that call for the epistemic resistance struggles of liberation movements in two parts.

2 See esp. the collection of essays edited by Paul Giladi and Nicola McMillan (2023).

3 See especially Lackey 2023 and Medina 2021.

In section 1, I will address worries about bringing together recognition theory and epistemic justice in a unified theory of social critique, developing an argument in defense of such unified theory. In section 2, I will argue that a unified theory of social critique needs to be grounded in grassroot movements of liberation. Focusing on the problem of alienation and self-ignorance, the argument of this section will be that properly diagnosing the epistemic injustices against oppressed groups can only be done from the engaged perspective of an activist, and that a critical consciousness about epistemic misrecognition (including *self*-misrecognition) can only be developed in and through the practices of consciousness-raising, collective learning and social transformation that liberation movements make possible. On my view, the kind of immanent critique of social pathologies of epistemic misrecognition that can be found in grassroot movements of liberation can avoid problems of paternalism and heteronomy.

1. Epistemic injustice and the Social Critique of Recognition Failures

In "Two Interpretations of Social Disrespect: A Comparison between Epistemic and Moral Recognition" (2023), Axel Honneth elucidates the strong convergence between his own theory of social recognition and Fricker's theory of epistemic injustice. Honneth emphasizes that what unites the two theories is the core idea that those who are subordinated or oppressed are not only materially disadvantaged, but also "mistreated with a certain condescension, disdain, and degradation that has to be conceived as unjust." (2023: 11) This form of mistreatment is explained in both theories as a form of *disrespect* (or lack of esteem) that results from the absence of social recognition or the presence of distorted recognition, that is, from being *misrecognized* as a moral and/or epistemic subject. While arguing for the priority of moral recognition over epistemic recognition, Honneth emphasizes that these two forms of recognition are intertwined in various ways: for example, being recognized as a moral subject requires (among other things) being recognized as being accountable for one's actions, which involves being recognized as capable of making sense of one's actions and of giving testimony about them. In a similar vein, Fricker also calls attention to the *hybrid* character of epistemic justice as involving the proper recognition of the moral and epistemic agency of subjects; and her entire account is framed as an ethics of knowing. Being properly recognized as a subject and participant in moral and epistemic practices is crucial for just treatment; and when such recognition is not given, there is a failure of justice that warrants struggles for recognition aimed at being counted as an equal. As Honneth puts it, "in my analysis, it is the practical subject, in Fricker's the epistemic subject, whose experiences of disrespect are made central to normative analysis." (2023: 14)

Following Hegelian insights, Honneth emphasizes the intersubjective nature of moral subjectivity: the moral subject becomes possible in and through relations of reciprocal recognition whereby one's status and agency as a subject of moral value are affirmed. Similarly, the epistemic recognition of others is crucial for the constitution of epistemic subjectivity, which is something that Fricker and her followers[4] have emphasized when they elucidate how the misrecognition of one's epistemic capacities and agency undermines one's status as a subject of knowledge and understanding and contributes to the deterioration of one's epistemic character and subjectivity—think, for example, of how difficult it is to maintain self-trust when one is systematically distrusted by others. As Honneth puts it, "the moral subject [...], due to her intersubjective nature, can establish a functioning *self-relation* only if she and her various capacities are adequately recognised by her fellows." (2023: 13, my emphasis) And the same is shown to be true of the epistemic subject by Fricker's theory of epistemic injustice. Therefore, Honneth rightly concludes, moral recognition theory and the theory of epistemic injustice teach us how different "*kinds of disrespect violate the normative conditions of realising an unimpaired self-relation as either an epistemic or a moral subject*," (2023: 17) showing that misrecognizing people morally or epistemically amounts to "treating them as something less than a fully-fledged human being." (2023: 18)

How can the erosion of one's humanity resulting from these different forms of disrespect be protested and fought against? Drawing from Honneth's insights, let's look more closely at how recognition struggles can begin in the aftermath of moral and epistemic forms of misrecognition that, as Honneth puts it, impair one's "self-relation as either an epistemic or a moral subject." One can only recognize oneself as a moral and epistemic subject if one is also recognized as such by others; this "functioning self-relation" cannot be established otherwise. Moral and epistemic subjectivity requires relating to oneself in a particular way, namely, recognizing oneself as a moral and epistemic subject with certain capacities; but this self-recognition in turn requires being properly recognized by others, which is precisely what it not available to the misrecognized

4 See Medina (2013) and Lackey (2023).

subject. One cannot begin the struggle for recognition without having at least a *minimal* sense of oneself as a moral and epistemic subject, and therefore a sense of what one is owed and not given in the recognitional dynamics one finds oneself in. In other words, the subject struggling for recognition needs to be capable of self-recognition, not only in a generic sense of seeing themselves as a moral and epistemic subject, but also in the more specific sense of recognizing oneself morally and epistemically *in excess of* the recognition received from society and its institutions. This seems to create an important paradox, namely, the *paradox of resistant subjectivity*: subjects who lack proper recognition from others seem to be ill equipped to protest the misrecognition they suffer because a critical consciousness of their misrecognition requires recognizing themselves fully and properly and they do not have the required social support for such self-recognition. How can morally and epistemically injured subject develop the capacity to recognize themselves otherwise and embark on recognition struggles?

This paradox would be unsolvable if we expected isolated individuals to start struggles for recognition by themselves, in isolation and independently of each other. But, of course, this is not how recognition struggles begin, as a cursory look at the history of grassroot liberation movements can show us. In the next section, "Coming to Recognize Oneself (and Others) Properly," I will discuss how a critical consciousness about social pathologies of misrecognition can emerge in and through the shared activities developed by communities of resistance and grassroot movements of liberation, that is, how the members of such communities and social movements come to recognize recognition failures through protest actions, resistant practices, and activism. For now, as a preliminary response to the paradox of resistant subjectivity, it suffices to say that struggles for recognition are rendered possible by gaps and deviances that can be found in economies of recognition and are exploited by networks of eccentric subjects in resistant interactions. This brings together two important insights: an insight about the *communal* nature of recognition struggles, and an insight about the *polyphonic* nature of recognition dynamics.

In the first place, recognition struggles seeking moral and epistemic repair are not only intersubjective but *collective or communal*: dyads of persons struggling for recognition are inscribed in communities and networks of interaction in which subjects recognize each other differently and forge new relations of recognition among themselves, thus becoming capable of comparing and contrasting their nascent relations of reciprocal recognition with the consolidated forms of misrecognition they suffer. It is not the individual alone (or even a dyad of individuals) who develops resistance to misrecognition; resistant attitudes and actions are developed by an emerging network or community of those who begin to cultivate expanded or alternative forms of recognition.

In the second place, it is important to note that struggles for recognition are rendered possible by gaps and deviances that can always be found in economies of recognition because recognition dynamics are heterogeneous and intrinsically *polyphonic*: there is always (at least the possibility of) diversity and plurality in intersubjective relations of recognition. No matter how uniform recognitional practices may become and how dominant and hegemonic the standard norms of recognition may appear, recognition dynamics are never completely monolithic; they always admit variation across subject position and relations; they can be differently textured and layered, and their underlying normative principles of recognition can be contextualized in various ways. Therefore, social practices of recognition tend to have some degree of diversity and plurality, some heterogeneity within them; and this makes it possible that even the most marginalized and misrecognized subject may nonetheless find a modicum of recognition here and there, and can gather these various forms of piecemeal recognition received from a limited few and in limited contexts to stitch together a sense of self, a way of recognizing oneself that departs from the perversive forms of misrecognition that they tend to be subject to. Eccentric subjects who deviate from dominant recognitional dynamics can start to develop a critical consciousness about the misrecognition they receive from mainstream society and its institutions, as they start to mobilize in a community of resistance that struggles for recognition.

I will return to these ideas and will elaborate them more fully in the next section when I discuss the role of collective or communal self-recognition in the recognition struggles of grassroot liberation movements. As I will argue in that next section, recognizing forms of misrecognition that amount to epistemic injustices requires protesting the social dynamics of recognition in ways that vindicate alternative forms of communal self-recognition and self-affirmation, which is what protest actions and protest movements teach us how to do. But before getting to the crucial role of grassroot social movements in diagnosing and resisting misrecognition, let me first elaborate a bit further the account of epistemic injustice as a social pathology of misrecognition that I have sketched in this section, following Honneth. I will do so by addressing two worries about bringing together recognition theory and considerations of justice that have been raised in the recent literature.

In *Diagnosing Social Pathologies* (2023) Frederick Neuhouser argues that normative assessments in terms of justice do not fare well when compared with theories of social pathology, which include theories of recognition such as Honneth's. More specifically, Neuhouser argues that theories of social pathology have two distinct advantages over theories of justice: an explanatory advantage and an evaluative advantage. According to Neuhouser, the first advantage that theories of social pathology exhibit over theories of justice is that the former are not only capable of identifying normative failures, but they are also capable of giving an account of the sociogenesis of those failures in terms of the dysfunctional social dynamics that create and perpetuate them. Whereas theories of justice in liberal political philosophy do not contain a social theory that explains the social production of normative failures and how to correct them, Neuhouser emphasizes that theories of social pathology explain how those failures originate in social dysfunctions that are hard to break and give guidance for resisting and overcoming those dysfunctions. The second advantage of theories of social pathology, according to Neuhouser, is that "they have at their disposal critical resources beyond those employed by most liberal political and social philosophy." (2023: 10) Whereas liberal theories of justice rely on formal and thin normative notions, theories that diagnose social pathologies, such as recognition theory, use *thick* normative notions for assessing "social life as spiritual," that is, "as informed by the aspiration of social members to unite in their social activity the ends of life with those of freedom." (2023: xiv) In other words, "a diagnosis of social pathology is always in part an ethical critique." (2023: 12) Theories of social pathology concern themselves with "failures in realizing the good, broadly construed, rather than in achieving the right." (2023: 11) They evaluate how the human flourishing or "spiritual" development of some members of society becomes stifled or truncated because of social dysfunctions, such as misrecognition. Neuhouser points out that considerations of justice are not irrelevant for these social theories, but he argues that these considerations do not go to the core of the ethical failures involved in social pathologies. So, for example, in Marx's critique of capitalism, Neuhouser points out, "the problem with alienated labor is not primarily that it is unjust, [...but rather,] that the conditions under which such labor is carried out make it impossible for laborers to realize spiritual goods—recognition, self-esteem," etc. (2023: 13)

Let's consider whether the worries that Neuhouser raises against theories of justice also apply to the theory of epistemic injustice. Does the theory of epistemic injustice also lack the crucial explanatory and evaluative advantages that

Neuhouser sees in theories of social pathology? Quite the contrary. I argue that what makes the theory of epistemic injustice distinctive and especially powerful is precisely the kinds of explanatory and evaluative advantages that Neuhouser highlights, which strengthens the case for its strong convergence with recognition theory and speaks in favor of using these two theories in tandem.

In the first place, what is so powerful about Miranda Fricker's (2007) theory of epistemic injustice is precisely that it does offer an explanation of how epistemic injustice results from social dysfunctions in credibility and intelligibility assessments. On her account, patterns of epistemic disrespect are grounded in an unfairly biased social imagination that mediates our interactions and epistemic appraisals of each other. Unlike the liberal theories of justice that Neuhouser focuses on, Fricker's theory of epistemic injustice does contain a social theory that explains the underlying mechanisms of the social dysfunction in question. Fricker's theory explains how epistemic misrecognition is grounded in a social imaginary that promotes prejudicial stereotyping. Honneth recognizes this and praises the explanatory power of Fricker's account. In fact, Honneth sees here a convergence between Fricker's theory of epistemic misrecognition and his own theory of moral misrecognition in that they both offer accounts of "the social mechanisms by which such forms of disrespect arise and become mental habits." (2023: 18) However, Honneth argues that Fricker's account of the sociogenesis of recognition failures in terms of prejudicial stereotyping is incomplete and needs to be supplemented. According to Honneth, without supplementation, Fricker's account would be insufficient because it focuses exclusively on cognitive and individualistic factors, and the cognitive explanation that Fricker offers needs to be supplemented with an account of the motivations and social interests driving the recognition dysfunctions in question, an account that also enables us to see more clearly the structural and institutional dimensions of these dysfunctions. Honneth argues that the efficacious and recalcitrant nature of the prejudicial stereotyping underlying misrecognition "stems from dominant groups' deep-seated need to find evidence supporting the putative rationale for their privileged social position." (2023: 20) Therefore, moral and epistemic disrespect "should be interpreted as resulting from a combination of 'ideas' and 'interests', to invoke Weber." (2023: 21) Honneth emphasizes that an account of misrecognition of this sort that combines cognitive and non-cognitive factors can explain how "such disrespect can assume an institutional, or indeed structural, character even within a (self-proclaimed) highly enlightened culture and a critical public sphere," for motivated prejudicial attitudes "seep into the modes of behaviour of administrative officials, public authorities, and companies, forming habits, altering the rules of social engagement and ultimately even finding themselves reflected in the architecture of buildings and interior design." (2023: 21) So, as supplemented by Honneth, there is no reason to think that a theory of epistemic injustice cannot share the explanatory power of theories of social pathology. In fact, the expanded account of the underlying mechanisms of misrecognition that Honneth proposes goes along well with recent expansions and supplementations of the paradigm of epistemic injustice that emphasize the structural and institutional dimensions of epistemic misrecognition.[5]

In the second place, Fricker's theory of epistemic injustice is clearly not constrained by a thin and formal notion of what is right that can be detached from thicker normative conceptions about full subjecthood and having one's humanity fully recognized and supported. So Neuhouser's worry about the thinness of considerations of justice in liberal political philosophy does not apply here, and Fricker's theory of epistemic injustice also enjoys the evaluative advantages that Neuhouser ascribes to theories of social pathology. Like these theories, Fricker's too offers an *ethical* critique. It is not accidental that the subtitle of Fricker's monograph is "power and the *ethics* of knowing"; and the core of Fricker's account of epistemic injustice can be understood as an ethical critique of social pathologies of epistemic misrecognition that are predicated on the ethical value of epistemic flourishing, that is, of having epistemic dignity and respect, having access to epistemic goods and agency in epistemic practices, and being supported in one's full development as a subject of knowledge and understanding. In his elucidation of the convergence between Fricker's theory and his recognition theory, Honneth emphasizes that both theories are concerned with the ethical failures involved in different kinds of disrespect that arise from patterns of misrecognition. However, Honneth argues that the virtue-theoretic approach to epistemic recognition and its failures that Fricker favors has its pitfalls. In particular, Honneth argues that Fricker's account of epistemic injustice as a virtue runs the risk of leading to "an overextension of individual responsibilities." (2023: 25) Given that epistemic disrespect should be thought of as resulting not only from attitudinal deficits but from "the institutional sedimentation of *interested ignorance*" (Honneth 2023: 24), responsibility for epistemic injustice should not be conceptualized primarily as a matter of the individual's duty to develop virtuous recognition of others, but rather, as a collective and institutional responsibility for changing practices

5 See esp. Medina (2021) and Lackey (2023). See also Samaržija and Cerovac (2021).

of recognition and their underlying norms. Here too there is a strong convergence between Honneth's proposal for amending Fricker's approach and recent expansions and supplementations of Fricker's approach that focus on the collective, structural, and institutional aspects of epistemic injustice (e.g. Medina 2021 and Lackey 2023).

If we don't treat issues of responsibility for recognition and its failures in the abstract, in terms of virtuous dispositions that all individuals must have, but rather in concrete socio-historical contexts, in terms of specific institutions and economies of recognition historically developed, then, as Honneth puts it, this "historical institutionalism [...] requires accentuating an element of social practices much more strongly than Fricker does." (2023: 29) It is through social practices that we became capable of diagnosing and resisting epistemic injustices. As we shall see in the next section, it is *only* through grassroot social practices that we can develop a critical consciousness of forms of misrecognition such as those involved in epistemic injustice, and it is *only* through grassroot social practices that we can mobilize to resist epistemic misrecognition and fight for more just recognitional dynamics. This is exactly how Honneth concludes his elucidation of the convergence between recognition theory and the theory of epistemic injustice:

> [If] one wants to explain how historically given recognitional orders can be superseded in the first place, one has to draw on the transformational power of a type of social conflict that I, following Hegel, have called a 'struggle for recognition'. This denotes a form of social protest and revolt whose primary motivational source is not an interest in material improvement, but social respect and recognition. (2023: 29)

2. Coming to Recognize Oneself (and Others) Properly
Coming to recognize recognition failures through protest actions, resistant practices, and activism

How can patterns of epistemic misrecognition be recognized? Neuhouser points out that social pathologies are often invisible and it takes a social movement to make them visible. As he puts it, "social suffering must be articulated and made comprehensible to those who do not suffer from it." (2023: 4) And Neuhouser goes on to observe that in the social movements of the twentieth century, such as the Civil Rights movement, the Women's movement, or the

Gay Liberation movement, "the sufferers" themselves saw as their "principal task [...] to articulate the meaning of their suffering to others." (2023: 4) This is an important point, but it captures only one part of the struggles for recognition led by social movements of liberation, for the central challenge of these movements is not only to make a social pathology of misrecognition visible and audible *to others*, but also *to themselves*. That is, the victims of epistemic misrecognition and disrespect also have to struggle to come to recognize the misrecognition that they suffer. In her primary example of epistemic injustice, Fricker emphasizes how hard it was for women to communicate their experiences of mistreatment at the workplace before they developed the language of sexual harassment; but she also emphasizes that it was extremely hard for women to properly understand these experiences *themselves*, and it took many "speak-outs" before they could properly recognize them. On the other hand, in her discussion of Edmund White's *A Boy's Own Story* (2007: 163ff), Fricker underscores that some cases of hermeneutical injustice can be "so damaging that it cramps the very development of self" (2007: 164). According to Fricker's analysis, White's autobiographical narrative illustrates how a gay subject in a homophobic environment is not only epistemically disrespected by others but also by himself, since he becomes incapable of trusting himself and his experiences of desire and, therefore, incapable of giving himself proper epistemic recognition. Fricker emphasizes here that the emergence of a self-capable of resisting this misrecognition (what I call a *resistant* subjectivity) requires a "psychological rebellion," for the subject would have to become able to "rebel against internalized yet falsifying hermeneutical constructions of one's social identity" (165) But how can this *inner rebellion* be achieved? Let's look more closely at the difficulties that internalized misrecognition poses for victims of misrecognition to embark upon struggles for proper recognition, thus returning to the paradox of resistant subjectivity formulated in the previous section.

That proper self-recognition under conditions of oppression is challenging and takes a struggle to achieve should not be surprising since, as I have argued elsewhere (2013), systems of oppression produce bodies of active ignorance that include not only ignorance about others but also ignorance about oneself. This self-ignorance takes a particular shape that is important for our purposes to understand, since it blocks the path to develop critical consciousness about misrecognition, that is, it undermines the struggle to come to recognize one's own misrecognition. The cultivation of bodies of interested active ignorance incentivized by a system of oppression to protect itself includes the promulgation of self-ignorance among the oppressed. But how is one blocked to know things about oneself? This is accomplished in two ways: by arranging social life in ways that alienate oppressed subjects from their own experiences and create forms of self-estrangement; and through ideological distortions that offer ready-made interpretations of the subject's experiences as something other than experiences of oppression, blocking the development of a critical consciousness about the misrecognition of oneself and one's experiences. One may think that if the subject does have these experiences available within themselves (e.g. women do experiences unwanted sexual attention as something other than "harmless flirting", gay individuals experience same-sex desire as something other than a "perversion", etc.), how is it that they become unable to develop an alternative understanding of their own experiences and unable to recognize themselves as being harmed by the dominant misrecognition? On my own view, the answer is that the subject's own *epistemic agency* is blocked or undermined so that they cannot develop their own alternative understanding and the alternative expressive and interpretive resources that they need to formulate that understanding. In my view, this fits perfectly well with what has been called *agential epistemic injustice* in the recent literature, but with the twist that the subject is recruited to block or undermine their own resistant epistemic agency, so I will call it *self-perpetrated agential epistemic injustice*.

There is a higher-level epistemic injustice that prevents individuals from being able to protest the epistemic injustice they suffer, and this higher-level epistemic injustice involves a *self-perpetrated agential epistemic injustice* in which the subject is recruited to undermine their own epistemic agency in attempting to understand and give testimony of aspects of themselves. Following Fricker's distinction, we can identify here two distinct types: there is *self-perpetrated agential hermeneutical injustice* when a subject blocks their own attempts to understand aspects of themselves or areas of their experiences because those attempts are rendered difficult and risky and any departure from dominant hermeneutical sensibilities is stigmatized; and there is *self-perpetrated agential testimonial injustice* in attempts to talk about aspects of themselves or areas of their experiences *even to themselves* (e.g. in diaries or acts of self-acknowledgements) or to people like them. These are different ways of unfairly blocking the oppressed subject's epistemic agency to even attempt to develop and share interpretations and assessments of their lives in their own terms, and this includes the self-blocking of one's own epistemic agency in such expressive and testimonial attempts. And note that this is different from testimonial or hermeneutical *smothering*[6] because it is not just about being inhibited by hostile environments and the prospect of defective uptake, but rather, by one's own inability for proper self-recognition.

Since, on my account, the problem consists in being *disempowered* to develop resistant epistemic agency and thus unable to begin the struggle toward proper recognition, the fight against self-perpetrated agential epistemic injustice will require *becoming empowered*. But where can this empowerment come from? It can only come from oppressed subjects encountering each other, starting to recognize each other, and starting to support each other in their shared struggles. This is why struggles for recognition need to begin with *practices of self-empowerment*, practices in and through which oppressed subjects empower each other so that they can, together, get out of the traps of misrecognition. Only in this way can the oppressed subject get out of the predicament of self-blocking that prevents resisting misrecognition. As briefly explained in the previous section, my solution to the paradox of resistant subjectivity resides in a polyphonic and communal view of recognition dynamics and of the resistance against the normative failures in such dynamics. Grassroot movements of liberation tap into this polyphony when eccentric voices encounter each other and they start a social mobilization to speak together against their misrecognition. As I have discussed in detail in *The Epistemology of Protest* (2023), the activities of consciousness-raising that we find in liberation movements such as the Women's movement or the Queer Liberation movement are ways of resisting the self-misrecognition that prevents victims of injustice from seeing themselves as having been wronged in the first place, supporting individuals to overcome the self-blocking that prevents them from seeking recognition. Becoming capable of a kind of self-recognition required for resistance (recognizing the misrecognition of one's own worth) needs to be supported by the recognition of others. The self-recognition of victims of injustice and the mutual recognition of fellow victims go hand in hand and depend on one another; these interdependent forms of recognition are used in liberation movements for creating and sustaining communities of resistance. The visibility actions or pride actions of the Queer Liberation movement, such as kiss-ins, offer a paradigmatic example of this, as I have discussed elsewhere (2023 and Forthcoming).

Individual members of an oppressed group can muster the courage and develop the agency to resist misrecognition *not by themselves and in isolation*, but

---

6 See Dotson (2011).

when they come out, encounter each other, and participate in shared resistance struggles where they start developing alternative forms of recognition that make it possible for them to establish self-recognition as they are being recognized by each other. Shared activities of resistance, such as the speak-outs or visibility actions that we find in liberation movements, offer not only the opportunity, but also the motivation, encouragement, and empowerment that oppressed subjects need to overcome the self-perpetrating agential epistemic injustices that prevent them from resisting misrecognition and seeking alternative forms of recognition. These shared activities of resistance are practices of *self-empowerment* that are directed primarily at the oppressed subject themselves, not at outgroup members: they aim at empowering those who share the predicament of being mistreated and being disempowered to protest their mistreatment. Self-recognition is a crucial part of the struggle of social movements that need to focus, first of all, on social mobilization, that is, on supporting and facilitating the emergence of a critical consciousness, on community-building, and on the creation of a public that didn't exist before. Unlike Neuhouser (2023), who focuses exclusively on the role that social movements play in making misrecognition visible "to others", Honneth (2023) acknowledges that the recognition struggles of these movements start with *an inner rebellion* and, before they reach out to wider publics, they focus on building a counter-community, cultivating alternative forms of recognition, and developing alternative languages and interpretative resources.

Honneth rightly points out that "such struggles generally begin in small-scale domains: in private households, places of work, on public transport or in administrative offices, places where isolated individuals contend with representatives of power holders over the meaning and normative horizon of established recognitional norms." (2023: 29) Indeed, grassroot social movements of liberation start with micro-resistance in everyday life, with micro-struggles for recognition in ordinary interactions; and, building on those micro-practices of resistance, they develop larger-scale campaigns of mobilization for proper recognition. In this sense, it is not difficult to see the importance and subversive potential of the everyday micro-struggles for self-recognition that started the Queer Liberation movement: everyday interactions among queer subjects who made themselves visible to each other, practices in which queer individuals expressed pride to each other and reclaimed their own alternative spaces outside the mainstream, spaces—ranging from private or semiprivate (communal gay houses) to non-mainstream public spaces (such as gay bars)—in which queer visibility and queer pride became possible. These small-scale, routine practices of resistance paved the way for the visibility actions and pride actions that the Queer Liberation movement used in national campaigns like Queer Nights Out, carried out by Queer Nation in the 1990's. When activists of Queer Nation started their queer visibility campaign and staged kiss-ins in street corners and straight bars, it was definitely not the first time that queer subjects kissed in public and expressed pride in their sexual behavior. It was because they had been able to do that first in private and semiprivate spaces as well as in non-mainstream public spaces (gay baths and gay bars, for example), because they had empowered each other and developed expressive and interpretive agency to make sense of themselves in a different way, that they were able, eventually, to defy the mainstream and present their behavior to the world as worthy of recognition and pride.

The defiance that characterizes the struggle for recognition of the Queer Liberation movement was slowly developed by defiantly resisting misrecognition together in everyday spaces and interactions, by supporting each other in the refusal to hide and to accept shame and invisibility, by developing micro-practices of queer pride and self-empowerment. Self-recognition and self-empowerment are crucial in the initial stage of a grassroot liberation movement because they are necessary to get the recognition struggles started; but they also remain central in later stages of liberation movements and in their public interventions to create new recognitional dynamics. We can see this clearly in the public visibility and pride actions of Queer Nation in the 1990's. Let's look at their kiss-ins more closely to see how queer activists aimed at self-recognition and self-empowerment in these public acts of protest.

Through defiant anti-stigmatization protests, in the 1990's Queer Nation fought for the dignity and proper recognition (ethical, political, and epistemic) of queer subjects through the expression of pride in sexual difference, alternative community-building, and practices of queer self-empowerment. Queer Nation encouraged queer subjects to feel unapologetic and proud of their sexual lifestyles and of their alternative communities and countercultures. The inaugural protest act of Queer Nation was a kiss-in that was staged as a visibility action at Flutie's, a New York straight bar, on April 13, 1990. That night dozens of same-sex couples entered Flutie's and started making out, deeply disrupting the heteronormative expectations that had structured that space up to that point, to the discomfort of the owner and many (if not most) of the establishment's regular clients. Visibility protest actions of this sort sponsored by Queer Nation came to be known as "Queer Nights Out" and became popular in some metropolitan areas in the 1990's. Probably the most disruptive and best publicized kiss-in of this kind was organized by Queer Nation/L.A. in 1991 to interrupt the 64th Academy Awards by obstructing entry to the event with a multitude of same-sex couples kissing on the red carpet. Mainstream sensibilities felt insulted and disrespected: "Why are these queers rubbing their sexual preferences against people's noses in public? Why are they flaunting their indecent attitudes and behavior in everyone's face?" Queer Nation's kiss-ins were visibility actions that tried to de-stigmatize queer identities and deeply transform their social (in)visibility, turning the invisible and shameful into something visible and honorable, the object of pride. As I have argued elsewhere (Forthcoming), these kiss-ins contained both in-group and out-group communication with a deeply subversive potential; they aimed to achieve two things in the struggle for recognition: one is negative, reactive and deconstructive, while the other one is positive, proactive, and constructive. On the negative, reactive and deconstructive side, queer kiss-ins offered a public critique of and resistance against social stigmatization: in kiss-ins, protesters denounced and resisted the denigration and exclusion of their sexual identities. On the positive, proactive, and constructive side, Queer Nation's kiss-ins queered the public sphere and defiantly disrupted the heteronormative expectations of dominant sensibilities in order to assert queer dignity and demand proper recognition, creating new possibilities for social interaction and prefiguring more inclusive possible futures.

As I have argued elsewhere (Forthcoming), the central goal of Queer Nation's uncivil activism was not so much to persuade current mainstream publics, but to make room for the meaningful expression of the alternative sensibility of queer subjects and their allies. A central critical aim was *empowering an alternative sensibility*: the sensibility of queer counter-publics. What the uncivil activism of Queer Nation seems to have been most interested in was the transformative effects that aggressively confronting the mainstream public could have, not for members of that public, but rather, *for themselves*, for queer subjects bonding together and empowering themselves, while garnering the solidarity and support of at least some (no matter how small) cross-section of the American public. This puts *self-empowerment* at the center of the communicative dynamics of their protest acts: their primary audience was the *internal* audience of queer subjects and their allies; the primary goal was to develop an alternative sensibility for the queer community that could support self-recognition, a proud and defiant sensibility that could overcome the feelings of shame and other disabling negative emotions rooted in experiences of stigmatization and social rejection.

Queer Nation's uncivil protests against homophobic taboos focused on the self-affirmation and self-empowerment of a visible queer public. This is brilliantly expressed in a slogan used in one of Queer Nation's T-shirts: "I am out, therefore I am." Queer Nationals encouraged queer subjects to come out and to feel like they existed in the public sphere: because I am out, I exist as a queer person in the public sphere; and because we are out together protesting in public, we exist as a collective subject that can stand up and speak up for itself. Queer acts of coming out were acts of protest that transformed the public sphere: acts that created and affirmed public standing for queer subjects and, thus, interrupted the hegemonic control of heteronormativity over the public sphere. The importance of creating a counter-community and a counter-culture in which stigmatized groups can find recognition cannot be overemphasized. The making of a counter-community and a counter-culture involves the development of an alternative normative economy of recognition that does not depend on the norms and forms of recognition and respect of the dominant, mainstream world. Because of this subversive and transformative potential, Marx, neo-Marxist philosophers and recent critical theorists have underscored the importance of counter-cultures for overcoming social exclusion and oppression. Honneth, for example, has argued for a politics of collective self-affirmation through the independent recognition of oppressed subjects among themselves. Honneth (2012) describes the alternative normative economies of recognition developed by disenfranchised groups as "countercultures of respect"; and he argues that the alternative forms of recognition developed by these countercultures are *compensatory*: they compensate for the recognition deficits accrued by subjects who have been oppressed and deprived of dignity and respect. I hope to have shown how important practices of self-empowerment, as exemplified by the visibility actions of Queer Nation, are for recognition struggles: it is only through grassroot practices of this sort that the recognition struggles of liberation movements can get started, fighting together on the grounds the blocking of the emergence of resistant subjectivities and facilitating the self-recognition of harmful misrecognition.

Grassroot movements of liberation enable their members to develop critical consciousness about their misrecognition despite the pervasive forms of alienation and ideological distortion that systems of oppression put in place. Through their recognition struggles, grassroot liberation movements make it possible to develop a social critique of misrecognition that is both *immanent and transcendent*: it is immanent because it emerges from *inside* contexts of oppression, from the very critical experiences of those who suffer injustices in those contexts and start stitching their critical experiences of injustice together into a new sensibility; but it is also transcendent because it points to something *beyond* that context, something that is prefigured by the alternative sensibility and counter-culture cultivated by the grassroot practices of the liberation movement, an alternative form of recognition. The subject position of the resistant subjectivities that liberation movements make possible can be described as an *insider-outsider* perspective: the critical perspective of those who, while being insiders in a given social context, have nonetheless been excluded from that context and forced to recognize themselves outside that context, that is, at the margins or in interstitial spaces that they have turned into a new context for their own self-recognition and self-affirmation.

Insider-outsiders are what Patricia Hill Collins (2000) aptly describes as "outsiders within," that is, subjects who participate in a practice while being treated as strangers who don't belong, which gives them distinctive epistemic advantages and a critical edge: they are able to notice things that others don't and to develop a critical perspective from their experiences of not belonging or not being fully accepted. It is by virtue of being insiders who do not fully fit in that these subjects are capable of identifying normative deficits, such as recognition failures, which go unnoticed by mainstream or dominantly situated subjects. Through the engaged perspective of activists who are insider-outsiders, liberation movements are able to develop *transformative immanent critiques* of social pathologies of recognition. This kind of immanent critique[7] is neither internal reformism nor external interventionism: it is a critique that does not remain (entirely) within the insider's perspective and is capable of transcending that perspective; but, at the same time, it is a critique that does not come (entirely) from the outside and does not indulge in paternalistic interventions that violate the autonomy of a community or a social practice. The kind of immanent critique of social pathologies of misrecognition developed by grassroot movements of liberation can avoid problems of paternalism and heteronomy: this critique does not come from an external normative standpoint that self-proclaims to know better and dictates how a recognition dynamics (or "recognition order," to use Honneth's terminology) should be meliorated; rather, it is the subjects themselves experiencing the problems of those dynamics that develop the critical sensibility to diagnose social pathologies of recognition and work toward repairing the harms and toward the cultivation of proper and just recognition.

7 For a compelling account of immanent critique of this kind, see Jaeggi (2018).

The grassroot approach to the diagnosis of social pathologies of recognition can thus solve the problem of epistemic and normative authority that arises for critical social theories. The therapeutic and diagnostic language of these critical social theories invites a question about epistemic and normative authority: who is the doctor who can diagnose a social pathology and prescribe a cure? Who has the authority and expertise to identify social illness and the path to social health? Who can be the subject and agent of the kind of social change that can bring about social healing? The answer of the grassroot approach is: the people themselves; in particular, those who suffer social exclusions and can develop critical insights and critical consciousness by being in touch with and transformed by their experiences of marginalization and oppression; they are the ones who have understanding and knowledge of social exclusions and insights into what needs to change for the social fabric to heal, become more inclusive, and allow for new paths of social communication and social growth. The diagnosis of social pathologies of recognition and the fight against them are not driven by an external standpoint or a *deus ex machina*. According to the grassroot approach I have defended in this essay, struggles for recognition have to be grounded in the critical experiences of those who suffer from recognition failures and injustices (such as patterns of epistemic misrecognition and the resulting epistemic injustices); only those insider-outsiders who develop a critical consciousness about misrecognition can guide the struggle of resistance against recognition failures and the fight for building or re-building a community that can create new dynamics and normative orders of recognition.

3. Recapitulation

In the first part of the essay, I have argued that epistemic injustices can and should be conceptualized as social pathologies of misrecognition, elaborating further Honneth's suggestions in this direction and addressing some concerns about bringing together recognition theory and the theory of epistemic injustice. I have further argued that resisting epistemic injustices involve recognition struggles, and that these struggles are hard to get started under conditions of oppression because oppressed subjects themselves are encouraged to misrecognize themselves and tend to be blocked from developing a critical consciousness about their own misrecognition (what I called *the paradox of resistant subjectivity*). In the second part of the essay, I argued that shared activities of resistance are required for overcoming the self-misrecognition of oppressed subjects (a form of *self-perpetrating agential epistemic injustice*), and getting recognition struggles started. These shared activities of resistance are *communal counter-practices of epistemic self-empowerment*, that is, practices in which those who have been epistemically disempowered to the point that they cannot even recognize aspects of themselves or areas of their experience start empowering each other to think differently and speak differently, in their own terms. I analyzed visibility actions (such as kiss-ins) of the Queer Liberation movement as a paradigmatic example of a practice of self-empowerment necessary for recognition struggles to get off the ground. Focusing on the problem of alienation and self-ignorance, I argued that properly diagnosing the epistemic injustices against oppressed groups can only be done from the engaged perspective of an activist (the *insider-outsider*), and that a critical consciousness about epistemic misrecognition (including *self*-misrecognition) can only be developed in and through the practices of consciousness-raising, collective learning and social transformation that liberation movements make possible. Grassroot liberation movements are capable of developing an *immanent critique* of recognition dynamics that aims at the self-transformation of those dynamics from the inside, avoiding the strictures and problems of both internal reformism and external interventionism.

References

Dotson, Kristie (2011): "Tracking Epistemic Violence, Tracking Practices of Silencing," Hypatia: A Journal of Feminist Philosophy 26 (2):236-257.

Fricker, Miranda (2007): *Epistemic Injustice: Power and the Ethics of Knowing*. New York: Oxford University Press.

Hill Collins, Patricia (2000): *Black Feminist Thought: Knowledge, Consciousness, and the Politics of Empowerment*. Boston: Unwin Hyman.

Honneth, Axel (2012): *The I in We: Studies in the Theory of Recognition*. London: Polity.

Honneth, Axel (2023): "Two Interpretations of Social Disrespect: A Comparison between Epistemic and Moral Recognition," in Giladi and McMillan (eds.).

Giladi, Paul, and McMillan, Nicola (eds.) (2023): *Epistemic Injustice and the Philosophy of Recognition*. New York: Routledge.

Jaeggi, Rahel (2018): *Critique of Forms of Life*. Cambridge, MA: Harvard University Press.

Lackey, Jennifer (2023): *Criminal Testimonial Injustice*. New York: Oxford University Press.

Medina, José (Forthcoming): "Breaking Silences: Public Kissing and Other Things to Do When Protest Must Be Uncivil," in Candice Delmas and Avia Pasternak (eds.), *Uncivil Resistance*. New York: Oxford University Press.

Medina, José (In Preparation): "Protest Movements and the Right to Make Oneself Known."

Medina, José (2023): *The Epistemology of Protest: Silencing, Epistemic Activism, and the Communicative Life of Resistance*. New York: Oxford University Press.

Medina, José (2021): "Agential Epistemic Injustice and Collective Epistemic Resistance in the Criminal Justice System," *Social Epistemology* 35: 2, Special Issue: *Epistemic Injustice and Collective Wrongdoing*, 185–196.

Medina, José (2013): *The Epistemology of Resistance: Gender and Racial Oppression, Epistemic Injustice, and Resistant Imaginations*. New York: Oxford University Press.

Moody-Adams, Michele (2022): *Making Space for Justice: Social Movements, Collective Imagination, and Political Hope*. New York: Columbia University Press.

Neuhouser, Frederick (2023): *Diagnosing Social Pathologies*. New York: Cambridge University Press.

Pohlhaus, Gaile (2020): "Epistemic Agency under Oppression," *Philosophical Papers* 49 (2): 233–251

Samaržija, Hana, and Cerovac, Ivan (2021): "The Institutional Preconditions of Epistemic Justice," *Social Epistemology* 35 (6): 621–635.

Class as Moral Injury

Jacob Blumenfeld

Everyone today denies the existence of class, thus proving its enduring power. Class, it is said, has disappeared as a political category, as an economic indicator, as a cultural marker and as a specific identity. Classes have been integrated, recuperated, coopted, in short: *surpassed*. And yet economic inequality is skyrocketing, rents in the world's major cities are unaffordable, social programs are being cut, people are working more hours for less pay, and prices are rising faster than wages. Nevertheless, union membership is declining, political parties rarely support working class policies, and neoliberalism seems more triumphant with every crisis. As some Friends put it, "everywhere proletarianized individuals, nowhere the proletariat" (Friends of the Classless Society 2007). To put it sharply, if class has disappeared, then why is it still being so ruthlessly attacked?

There is a simple answer to this question: it is not class that has disappeared, but consciousness of it. In other words, what has vanished is any political power tied to the category of class. For some, this fact disproves the classical Marxist assumption that capitalism itself organizes the working class into a political power. Since class is no longer a meaningful pole around which people organize their social or political self-understanding, the argument goes, class theory itself must be false. Culture, status, ethnicity and identity are now the most important sources of collective self-understanding and thus the starting point for political change.

But this "farewell to the working class" story was never really accurate (Gorz 2007). On the one hand, capitalism never organized workers along class lines, but always in political and social *processes* that went hand in hand with *proletarianization* (Thompson 1964; Eiden-Offe 2023). On the other hand, class is not just an identity, but a socially determined position based on social wealth and power, whether one is aware of it or not. However, given this objective position, class can be *experienced* (or non-experienced) in very different ways. Understanding the specific mediation of class requires much more than an objective assessment of income levels.

I thus take class seriously as a meaningful part of our social life, even in times of its supposed eclipse. But I do so in a negative way, because it is not the positive identity of class that needs to be defended and explained, but the negative experience of class as a *moral injury to the status of being human*. This does not mean denying class as an objective-structural position, but understanding and describing class as an experience of suffering. It is thus more about class from the first-person perspective, not from the third person. Class should not be understood *as identity*, as this already presupposes the normative content of class as something positive. Rather, class should be treated as a non-identity, as an experience of lack or a lack of experience.

In what follows, I first discuss my understanding of class as structural vulnerability, i.e. the experience of being subjected to arbitrary power. I then discuss class as a kind of injury, both physical and moral. Using the short story of factory worker and autodidact Paul Mattick, I show how class mutilates human beings in both their bodies and souls (Mattick 1924; Endnotes 2024). I then draw out some conceptual consequences of viewing class as a specifically moral injury. Finally, I deduce what a negative theory of class might mean and what options exist for dealing with class as a moral injury. I argue that recognizing class as a moral injury means acknowledging the immanent demand to overcome it.

1. Class and Misrecognition

What is class? Class can be understood as a passive state or as an active process, as a sociological category based on income, or as a Marxist concept based on one's relation to the means of production (Eiden-Offe 2023: 5–10; Wright 2023). Class can be conceived, for example, through a theory of habitus or a theory of conflict, through statistical measures or through theories of recognition (Bourdieu 1984; Dahrendorf 1959). I argue however that class must be understood as an index of vulnerability to the economic power of others, i.e. to social domination.[1] Where one stands on the class spectrum depends on how

1 See Cicerchia 2021, 616–7: "Class is a condition of collective vulnerability among individuals to both shared constraints and to one another." For more recent debates on

vulnerable one is to being dominated and how much power one has to dominate others. To be dominated, in terms of *class*, is to be subject to the arbitrary power of personal and impersonal sources that determine access to the means of social reproduction. Control and ownership of social wealth form the class structure, mediated by legal property relations. Class relations within this structure become antagonistic through competition both vertically (between capitalists and proletarians) and horizontally (between capitalists and capitalists or between proletarians and proletarians) (Cicerchia 2021: 617; Mau 2023: 123–142; Brenner 2007: 58).

When I speak of class, I am primarily referring to the "class structure" as a specific social structure that determines access to and control over labor and wealth as well as the ability to satisfy one's needs. In other words, class names the real objective structure that restricts one's own sphere of freedom insofar as one must behave in a certain way in order to reproduce oneself. In bourgeois society, this means acting according to market imperatives. This class relation can only be cushioned *by owning property* or *through collective action*.

When I speak of the *moral injury* of class, I refer to the subjective and intersubjective experience of wage-dependent proletarians, whether they work or not. I mean those whose class position renders them systematically vulnerable, insecure, precarious and unfree, as well as subject to disregard, disrespect and injury – both physical and moral. The owning class also operates within this structure, but its ownership of wealth or productive assets shields it from the vulnerability suffered by the wage-dependent, who are forced to adapt their will to heteronomous ends in order to successfully fulfill their tasks.[2]

As mentioned earlier, I take class as a condition of *systematic vulnerability* to the economic power of others who may be owners of land, managers of capital or providers of wage labor. So what is the wrong of class? The usual answers run along the lines of theories of *inequality*, *exploitation* and *injustice* (Cicerchia 2021). These perspectives assume that the harm of class can be mitigated by balancing class relations between those who live off income and those who live off

class theory, see Mattick 2018, 191–218; Mau 2023, Chibber 2022, McCarthy and Desan 2023, Roberts 2023.

2 There is another specific psychological use of the term "moral injury", similar to PTSD, which refers to exposure to harmful events that transgress one's moral boundaries, for instance, in war. The philosophical meaning, although not always clearly defined, is different. I use the term moral injury here specifically to explore the inner damages of class to a person's sense of self.

wealth, reducing exploitative practices in the workplace, providing more benefits for workers, and making access to jobs, income, and wealth more equitable; the strategies that emerge from these theories are therefore concretized through labor law, income tax, equal opportunity policies, and social welfare. In short, they take the existence of classes as a given and, building on this, seek to reduce the harm caused by exclusion, discrimination or exploitation through processes of social recognition and legal protection.

A better world, however, is not a world of better classes, but a world *without* classes. It is not about having a certain class status recognized, but about overcoming classes. What does that say about class? It is not simply an unequal distribution of access to resources but, as I put it, a moral injury to one's sense of self through being subordinated collectively to the arbitrary power of the market. Class is an affront to one's status as a self-determined subject, as one worthy of respect and demanding recognition. It is an injury to the person who wants to contribute to society, but can only do so by destroying their own dignity and eroding their own character (Sennett and Cobb 1972).

Recognition theories and class theories face different challenges. Recognition theories often reduce class to an identity that should be recognized within a framework of progressive rights (Honneth 1995; 2024; Dejours et al. 2018). Class theories, on the other hand, criticize recognition as a superficial affirmation of identity that does not address the material conditions that underlie class structures (Fraser and Honneth 2003; Coulthard 2014). Yet, both theoretical perspectives are needed to understand how class can be experienced as a form of misrecognition, while at the same time demanding more than just recognition in order to improve it (Felski 2021; Scott 2022). In other words, the working class, characterized by vulnerability and dependency, cannot simply strive for recognition within the class structure without reinforcing the conditions of its own subordination. Therefore, the notion of class as moral injury aims at the abolition of class rather than its recognition.

As an existential condition of human life under capitalism, class shapes the self-image of individuals, their relations to others and the possibilities for collective self-determination (Eribon 2019; Louis 2017). Class is therefore not only an expression of inequality or exploitation, but a system of domination that can tear apart the moral fabric of human life. Its paradox consists in being a moral injury and at the same time being constitutive of the relationships that make modern subjects who they take themselves to be. Class both forms and deforms subjects in equal measure. The experience of submitting to the abstract imperatives of an economic structure that determines access to the means of social reproduction is both liberating and humiliating. It enables the ability to reproduce oneself and at the same time restricts the freedom to determine oneself.

Class is a moral injury insofar as it is fundamentally a form of misrecognition. And yet recognition is not the answer. Why is this so? Because the problem is not *classism*, i.e. some prejudice against workers as a class, but the condition of class itself. It cannot therefore be a question of morally recognizing the working class as dignified, but of abolishing the class structure that reproduces the domination of some over others. Recognition usually means recognizing one's own identity or status as worthy of respect. On the one hand, the struggles of the working class for labor rights are a paradigmatic case of a struggle for recognition. On the other hand, recognizing workers as a class does not remove the cause of the moral injury to human dignity that sparked the struggle for recognition in the first place. In contrast to other demands for status recognition, there is no equal reconciliation between the class of labor and the class of capital. For the aim of class struggle is not to remain a worker, but to escape from being a worker, to escape from subjugation to the personal-impersonal systemic constraints of the economy, to overcome them and finally to abolish them. The secret of proletarian class identity lies in the fact that it tends towards its own abolition and not towards its self-constitution (Eiden-Offe 2023:12). Class is thus both a paradigm for the theory of recognition and a refutation of it.

2. The Physical Injury of Class

Class inflicts both physical and moral injuries, institutionalizing relations of vulnerability and dependency in everyday economic practices. The wage-dependent class is separated from the means of reproduction and therefore vulnerable to systemic injuries by those who control access to their conditions of survival. These include physical injuries – such as accidents at work – and moral injuries resulting from the denial of autonomy. The physical toll of class is vividly illustrated in a 1924 short story by Paul Mattick called *The Conveyor Belt! A True Story from a Factory*, in which the death of a worker is portrayed as the result of the priority of profit over human life (Mattick 1924; Roth 2015; Endnotes 2024). In this short story, a factory worker named Müller is about to take his break from work, "three minutes to twelve", and is looking forward to lunch with his wife and children when the foreman asks him to repair one of the hissing engines. The foreman's decision to delay shutting down the engine for three minutes encapsulates the moral injury of class. What happens next is tragic: while he is standing on a ladder repairing the engine's flywheel, Müller's wife and children come in to have lunch with him. He smiles, makes a silly move, and falls into the engine's belt—getting brutally crushed in front of his family.

> How sad and infinitely cruel this all was. Why these accidents, why all this, why these three minutes, these three minutes to twelve? Couldn't the engine have been turned off earlier, a wretched three minutes earlier? Three minutes set against a sea of blood and tears. A human being—a man who smiled, who loved, and this woman—how she cared and how she loved, and these children's laughter—these little outstretched arms, calling out: "Daddy, Daddy!" All this outweighs the three minutes, more, outweighs the entire world. But at three minutes to twelve, it had been too early to turn off the engine (Mattick 1924).

To be subjected to the rule of working time means to be injured by class belonging: physically and morally. The body is crushed by three minutes that could not have been shortened, the worker is denied the human dignity that grants him autonomy and time for his family. *Why these accidents, why all this, why these three minutes,* writes Mattick. Because, the story ends, "a law governs here, a harsh law that knows nothing of the real world: profit! It neither hears the whimpering of the wife nor sees the tears of the children. It knows nothing of this Golgotha of labour" (ibid). This is the law of value, the law that subordinates human desires to the whims of an impersonal god of money, whose will is carried out by human beings. Here the worker's life and labor are subjugated to the rhythm of the machine, making his humanity a dispensable by-product of efficiency. The tragedy is not an anomaly, but is part of the structural logic of capital. The experience of class is this: loss of control, loss of dignity, loss of freedom, loss of self.

The role of physical injury is also explored in depth in Nate Holdren's *Injury Impoverished* (2020), which catalogs the unquantifiable costs of industrial accidents: the mutilation of bodies, the grief of families, the despair of communities (Holdren 2020: 19–52). But these injuries are not just physical, they are moral ones too, that is, avoidable, intentional, and done without regard for the dignity of the worker. The worker's body, maimed by machinery, becomes a site of alienation, where the promise of dignity through work turns into the brutal reality of mechanized exploitation. Holdren cites William Hard's 1910 account of class injuries, *Injured in the Course of Duty:*

> We cannot translate into dollars and cents the infinite torture, physical and mental, of America's 500,000 annual industrial accidents. We cannot capitalize the anguished leap of the workman's nerves under boiling metal. We cannot set a price upon the horror in the widow's heart when she carries to burial an oblong block of cold iron (ibid: 54).

What Holdren then documents is the process by which particular human stories of class injuries slowly become quantified into abstract statistics to be used by lawyers and economists seeking social reform. The initial outcry over working class injuries moves from seeking concrete recognition of moral suffering to accepting a form of justice based on monetary compensation.

Statistics from the International Labor Organization (ILO) show that the number of work-related deaths and injuries exceeds the number of deaths from wars and traffic accidents. Every year, an estimated 2.78 million workers worldwide die from work-related accidents and occupational diseases, while another 374 million workers suffer from non-fatal work-related accidents (ILO 2023; 2025). These international figures illustrate the extent of the physical damage that the normal functioning of capitalism causes to those suffering around the world, with the working class bearing the brunt of dangerous conditions. But the moral injury to class goes deeper than the physical injury itself; it lies in the systematic disregard for the life and dignity of the worker.

In an interview from 1916, disability activist Helen Keller explained her conversion to socialism by studying the industrial causes of seemingly natural misfortune:

> I had thought blindness a misfortune. Then I was appointed on a commission to investigate the conditions of the blind. For the first time I, who had thought blindness a misfortune beyond human control, found that too much of it was traceable to wrong industrial conditions, often caused by the selfishness and greed of employers. And the social evil contributed its share. I found that poverty drove women to a life of shame that ended in blindness (Keller 1916).

The fragmented body of the worker becomes a site of physical and moral vulnerability. This vulnerability is not accidental, but systemic and reflects the imperative to extract maximum value from labor. The reduction of the human being to instruments of production represents a profound moral failure that degrades both each individual worker and society as a whole. Yet this degradation goes hand in hand with a form of resilience: many workers adapt, persevere and are proud of their work. This resilience, however, serves to maintain the system that exploits them. When there is no other way to satisfy one's needs, work becomes both a cause and a cure for suffering. The fact that this experience is not unique, but plays out according to class position, means that it is not individual but structural. Class as a condition means that one has no choice but to return to the source of one's misery, be it in one place or another, for one boss or another, without the possibility of exit.

Our bodies are not immediately given as complete wholes but are products of a social achievement aimed at being treated and recognized as a person (Bernstein 2005: 315). As Helmuth Plessner argues, we both *have* a body and *are* a body, and the specifically *human* form of life is shaped by the struggle to navigate this ambiguity day in and day out (Plessner 2019: 271–2; Plessner 2020: 34–5). Physical injuries fragment our selves, undo us, alienate us, but not all physical injuries are moral injuries. If someone steps on my hand, it can be an accident, or it can be a way to demean my status. So why are class and related physical injuries and accidents moral injuries? Because, at a certain level of analysis, they are not accidents but the result of a systemic vulnerability to physical injury, and thus it is the structure of class dependency that leads to injury. If capitalism is, according to Søren Mau, "the institutionalisation of insecurity", then dealing with class injury means dealing with the fundamental insecurity of one's market-dependent existence (Mau 2023: 130).

3. The Moral Injury of Class

In their ethnographic analysis of working class subjectivity, Sennett and Cobb refer to the wounds that penetrate deeper than the flesh into the soul of the proletarian subject as "the hidden injuries of class" (Sennet and Cobb 1972). They examine how class undermines human dignity and autonomy in deep and existential ways. Working people experience shame, powerlessness and resentment, but paradoxically internalize the validity of the hierarchical system that humiliates them. The erosion of long-term values such as loyalty and commitment further corrodes individual character. For Sennett and Cobb, the question of class raises fundamental issues of freedom and dignity. To belong to the working class means not being able to determine one's own destiny, but to be determined by the arbitrary decisions of others. Only by submitting to the class structure is one free from becoming superfluous. The paradox of class, for Sennett and Cobb, is that workers may feel ashamed of their status and resentful of those who are more educated and above them in the social hierarchy, while at the same time believing that their own lower status is justified. The resentment towards those who look down on them is offset by the belief that such humiliation is justified by their own class position. If a person's ability is seen as a sign of their individual worth, then their lack of class mobility is proof of their own worthlessness (ibid: 53–109). Through a close analysis of workers' class subjectivity, the possibility of class transition is both desired and feared; for not to be a worker is to lose respect for oneself (ibid: 18–30). In a sense, there is no alternative to feeling injured, since the very levers of class transition such as education and knowledge further degrade workers' own dignity. To protect themselves from vulnerability to misrecognition and disrespect, to moral injury, according to Sennett and Cobb, workers can only cut social ties, retreat into the sphere of the private and sacrifice social life.

The double harm of class lies not only in the constant vulnerability to physical and moral injury, but also in the fact that one finds meaning in the very structures that harm them. Workers' pride in their work and their identification with the product of their labor reflects a form of recognition in which one's humanity is simultaneously affirmed and denied. This duality underscores the central paradox of class: the conditions of recognition are also the conditions of non-recognition (Dejours et al 2018; Jaeggi 2014). The moral injuries of class is not an accidental harm, but a condition of existence under capitalism. Because of their class position, wage-dependents are not only deprived of recognition, but also subjected to a form of misrecognition that reduces them to mere instruments of production. According to Sennett and Cobb, this misrecognition manifests itself in the stigmatization of manual labor, in the devaluation of workers as uneducated or unskilled, and in the pervasive sense of powerlessness that accompanies wage dependency. Yet, this condition is not passively endured, but actively reproduced through the internalization of class norms. Workers strive for the ideals that marginalize them and seek recognition in a system that systematically denies them their dignity. This paradox of recognition – in which the desire for dignity leads to deeper subjugation – is central to understanding the moral injuries of class. The workers' striving for respect and recognition, whether through hard work or social advancement, only reinforces the structures of domination. Class recognition becomes a double-edged sword: it offers dignity while reproducing the conditions of indignity.

In The Corrosion of Character, Sennett continues the examination of the injuries of class by following the lives of the children of the workers from the previous study (Sennet 1998). This time the analysis takes place in the 1990s, when the neoliberal economy has become dominant and with it the idea of flexibility of place, time, self and activity. The flexibility of proletarian children, as well as their internalized flexible relationship to values, norms and beliefs, further harms them as respectable subjects; it is hard to strive for recognition when one must constantly betray one's worth to meet the needs of the market. According to Sennet, the neoliberal economy prevents individual wage earners from developing a coherent narrative of self and identity, as they must constantly adapt their values and preferences to market pressures. This drive undermines the possibility of fixed values and meaningful experiences. Norms such as loyalty, purpose and commitment are obstacles when flexibility is necessary to remain competitive. The market creates an objective and subjective sense of permanent catastrophe: lives change and are thrown into disarray, not because of nature, but because of society itself. Disaster is internalized, instability normalized. Flexibility enables economic success while weakening one's character beyond repair.

4. What is a Moral Injury?

Sennett and Cobb's account of the hidden injuries of class, alongside other important sources such as the young Hegel, Mead and Marx, forms one of the inspirations for Axel Honneth's account of the struggle for recognition (Honneth 1982). For Honneth, the indignity of working class life, as described in his recent book *The Working Sovereign*, calls for new forms of recognition (Honneth 2024). The struggle for recognition is constitutive for social identity, especially for workers. The difference I want to emphasize in contrast to other critiques, such as Nancy Fraser's, is that the experience of the working class should be theorized not only as a struggle for recognition of class identity, but as a demand for abolition (Fraser and Honneth 2003). This can only be understood if we see class as a *limit* to freedom and not a condition of it.

Moral injuries are harmful because they deny recognition of a person's moral worth (Honneth 1997; Hampton 1999; Bernstein 2005, 2015; Congdon 2016; Renault 2017). These injuries take the form of physical harm, humiliation, or disregard of moral accountability. In the context of class, moral injury is not just an interpersonal wrong, but a systemic condition. For Honneth, moral injuries are interpersonal harms that deny recognition of some salient moral features of another person. Moral injuries are thus wrongs based in misrecognition, in harming one's normative self-conception through physical or verbal means. According to Honneth, "it is not solely the bodily pain as such, but the accompanying consciousness of not being recognized in one's own self-understanding that constitutes the condition for moral injury here" (Honneth 1997: 23). For something to be considered a moral offense and not just a mere misfortune, it must include "the aspect of recognition being withheld or denied" (ibid). That is to say, "the characteristic feature of moral injuries consists in persons being disregarded in aspects of their positive self-relation—whose intersubjective acknowledgment they are fundamentally dependent upon" (ibid: 24). For Honneth, the three paradigmatic forms of moral injury are physical harm, disregard for the person's moral responsibility and, finally, humiliation and disrespect. Regardless of the specific form, however, "the core of moral injuries is located in the refusal of recognition" (ibid: 25). This denial of recognition is fundamental, a result of the fact that human beings are constitutively intersubjective creatures who are vulnerable to harm because their self-identity is based on recognition from another person.[3]

Honneth argues that moral injuries arise from misrecognition when individuals are denied acknowledgment of their sense of self. For Honneth, recognition is not just a formal affirmation of identity, but a fundamental condition of self-realization. The denial of recognition through the repetition of the class structure thus represents a profound moral injury that corrodes one's normative self-conception. While Honneth's framework helps clarify the form that moral injury takes in regards to class, it fails to recognize the very source of the moral injury in class itself. While recognition is necessary, it cannot resolve the structural contradictions of class. To recognize class means confronting the conditions that make such recognition impossible.

Another account of moral injury comes from J.M. Bernstein, who links moral injury to the failure of recognition that sustains personal independence (Bernstein 2005; 2015). For workers, the class structure means constant vulnerability and dependence, leading to a profound alienation of the self. Bernstein's emphasis on the interdependence of physical and moral integrity underscores the systemic nature of class-related harm. The degradation of the body – through unsafe working conditions, inadequate health care and exploitative labor practices is inextricably linked to the moral degradation of workers. But Bernstein also points to a deeper paradox: the very intersubjective conditions

3 Honneth's argument here develops from a reading of the young Hegel. For a different perspective on the young Hegel and recognition, see Blumenfeld 2024, 170–246.

that degrade the self contain the possibility of solidarity and resistance. For Bernstein, the dignity of the person is precisely that which emerges through forms of recognition, through which the intact, self-moving body comes into being: the dignity of the self is the reflective articulation of the moral integrity of the body. Because we are bodies, injuries to our bodies are injuries to the self. We remain radically dependent on our social environment in order to maintain our personal independence. This dependence signals our constitutive vulnerability. The recognitive constitution of self-consciousness aims to demonstrate the internal, conceptual and empirical connection between bodily vulnerability and social dependence. For Bernstein, moral injuries arise through loss of trust, humiliation, emotional devastation and denial of our dependence on others. They manifest as a violations of another person's dignity through non-recognition, reduction of the self to the body, existential helplessness and loss of trust in the world.[4] "Moral injuries are injuries to one's standing or status that become actual through modes of physically treatment" (Bernstein 2015: 15). This criterion, in my account, can be applied not only to actions, but also to social conditions and relations, with the condition of class belonging being paramount.

The injury of class can also be understood as a form of *epistemic injustice* in which members of a dependent class are not recognized as credible knowers (Hänel 2024). Their knowledge of work, society and politics is disregarded, reinforcing their subjugation and devaluation. The epistemic marginalization of workers, for example, reflects a broader ideological framework that privileges the perspectives of the propertied class and silences or marks as non-intelligible the voices of those who produce social wealth. This marginalization is not complete, however, as workers develop their own insights rooted in lived experience that can challenge dominant narratives. This tension between marginalization and resistance is central to the epistemic dimensions of class.

The moral injury of class cannot be remedied by recognition alone. Unlike identity-based demands which reaffirm group belonging, the wrong of class points to the need for abolishing the source of harm, that is, class belonging itself. Indeed, recognition within the existing framework risks reinforcing the very conditions it seeks to address. The working class is often misunderstood as

4 See Bernstein 2015, 15: "Loss of trust in the world occurs through the recognition that I am absolutely dependent on the other for my standing as a self or person, even for myself, and hence that my existential helplessness is not a mere potentiality, but a present and now ever-present actuality. I am devastated."

pitiable, less intelligent or ideological, and this view reflects middle-class meritocratic ideologies that fail to capture the systemic constraints that drive workers' actions. Rights-based recognition, while valuable, often formalizes mutual indifference and abstracts from the particularity that sustains misrecognition. Yet ideology also contains contradictions: the narratives that justify class domination also reveal its fragility. By exposing these contradictions, dominated classes challenge the normative foundations of their own subjugation.

Class is not only a descriptive but a normative category: it denotes the moral wrong of being collectively subject to vulnerability of domination. Workers in particular are subject to the mute compulsion of economic relations and conform to norms that undermine their dignity and autonomy. Those who are not wage dependent are still forced to bend their own will to abstract economic imperatives in order to succeed. But they have managed to cushion or shield their vulnerability through property and thus act as bearers of class rule rather than its subjects. The wrong of class here lies in its systemic domination, not just in its inequality or exploitation. However, this domination is never complete, but is always contested, both materially and ideologically, and this very contestation points to the possibility of overcoming class, even within the constraints of the present.

5. Conclusion: From adaptation to abolition

Rectifying the moral injury of class demands moving beyond individual adaptation and incremental reform. Strategies such as acquiring property, exercising voice, or demonstrating loyalty may offer temporary relief, but they miss the root cause: the class structure itself. Following Albert Hirschman's framework, individuals can respond to class domination through exit (striving for upward mobility), voice (advocating for change) or loyalty (adapting to existing conditions) (Hirschmann 1970). However, these reactions are restricted by systemic forces that structurally limit the individual's ability to act. Exit through upward mobility, for example, often means complicity with the very system that perpetuates class-based harm; while voice and loyalty are often undermined by capitalist pressures themselves. Nevertheless, these strategies – while limited – demonstrate the potential for collective action. The failure of individual action points to the need for solidarity. A response adequate to the magnitude of the harm would entail collective action of solidarity with the aim of abolishing class society through a form of transformative socialization

(Blumenfeld 2023; 2025). This presupposes strengthening class power in order to end class itself, like a game that would abolish its own rule (Theorie Communiste 2011: 132). Solidarity, both as a principle and as a practice, challenges the atomizing tendencies of capitalist forms of life and reclaims the collective agency necessary for systemic change. Yet solidarity itself is fraught with contradictions, as it must navigate the tensions between individual autonomy and collective responsibility, between immediate demands and long-term goals (DuFord 2022). However, these tensions do not weaken solidarity, but rather enrich its potential as a transformative force. Class as a moral injury is a systemic affront to dignity, autonomy and self-determination. It cannot be eliminated through recognition alone, but requires the abolition of class itself. This paper calls for a paradigm shift from recognition to abolition, emphasizing collective action and social transformation as the path to overcoming the moral injuries of class. This shift is not straightforward, for it requires confronting the paradoxes and contradictions that define class itself. And it forces us to recognize that the struggle for human dignity is inseparable from the struggle to abolish the conditions that block that dignity in the first place.

References

Bernstein, J.M. (2005): Suffering Injustice, in: International Journal of Philosophical Studies, 13.3, 303–324.

Bernstein, J.M. (2015): Torture and Dignity, Chicago: University of Chicago Press.

Blumenfeld, J. (2023): Expropriation of the expropriators, in: Philosophy and Social Criticism, 49.4, 431–447.

Blumenfeld, J. (2024): The Concept of Property in Kant, Fichte, and Hegel, New York: Routledge.

Blumenfeld, J. (2025): The Socialization Debate, in: Graduate Faculty Philosophy Journal, 46.1, 93–122.

Bourdieu, P. (1984): Distinction, Cambridge.

Brenner, R. (2007): Property and Progress, in: Wickham, C. (ed.), Marxist History-Writing for the Twenty-First Century, Oxford, 49–111.

Chibber, V. (2022): The Class Matrix, Cambridge: Harvard University Press.

Cicerchia, L. (2021): Why Does Class Matter? In: Social Theory and Practice, 47.4, 603–627.

Congdon, M. (2016): Wronged beyond Words, in: Philosophy & Social Criticism, 42.8, 815–834.
Coulthard, G. (2014): Red Skin, White Masks, Minneapolis: University of Minnesota Press.
Dahrendorf, R. (1959): Class and Class Conflict in Industrial Society, Stanford: Stanford University Press.
Dejours, C., et al. (2018): The Return of Work in Critical Theory, New York: Columbia University Press.
DuFord, R. (2022): Solidarity in Conflict, Stanford: Stanford University Press.
Eiden-Offe, P. (2023): The Poetry of Class, Leiden: Brill.
Endnotes (2024): The Young Mattick, https://endnotes.org.uk/dossiers/the-young-mattick.
Eribon, D. (2019): Returning to Reims, London: Penguin.
Felski, R. (2021): Recognizing Class, in: New Literary History 52.1, 95–117.
Fraser, N. and Honneth, A. (2003): Redistribution or Recognition? London: Verso.
Friends of the Classless Society (2007): 28 Theses on Class Society, in: Kosmoprolet 1, https://kosmoprolet.org/en/28-theses-class-society
Gorz, A. (1982): Farewell to the Working Class, London: Pluto.
Hänel, H. (2024): Epistemische Ungerechtigkeiten, Berlin/Boston: De Gruyter Brill.
Hampton, J. (1999): Defining Wrong and Defining Rape, in: Burgess-Jackson, K. (ed.), A Most Detestable Crime, New York: Oxford University Press.
Hard, W. (1910): Injured in the Course of Duty. New York: The Ridgway company.
Hirschman, A. (1970): Exit, Voice, and Loyalty, Cambridge: Harvard University Press.
Holdren, N. (2020): Injury Impoverished, Cambridge: Cambridge University Press.
Honneth, A. (1982): Moral Consciousness and Class Domination: Some Problems in the Analysis of Hidden Morality, in: Praxis International, 2.1, 12–25.
Honneth, A. (1995): The Struggle for Recognition, Cambridge: MIT Press.
Honneth, A. (1997): Recognition and Moral Obligation, in: Social Research, 64.1, 16–35.
Honneth, A. (2024): The Working Sovereign, Cambridge: Polity.
International Labour Organization (2023): A call for safer and healthier working environments, https://www.ilo.org/publications/call-safer-and-healthier-working-environments

International Labour Organization (2025): Statistics on safety and health at work, https://ilostat.ilo.org/topics/safety-and-health-at-work/

Jaeggi, R. (2014): Alienation, New York: Columbia University Press.

Keller, H. (1916), Why I Became an IWW, https://www.marxists.org/reference/archive/keller-helen/works/1910s/16_01_16.htm

Louis, E. (2017): The End of Eddy, London: Harvill Secker.

Mattick, P. (1924): The Conveyor Belt!, https://endnotes.org.uk/posts/paul-mattick-the-conveyor-belt-1924

Mattick, P. (2018): Theory as Critique, Leiden: Brill.

Mau, S. (2023): Mute Compulsion, London: Verso.

McCarthy, M. and Desan, M. H. (2023): The Problem of Class Abstractionism, in: Sociological Theory, 41.1, 3–26.

Plessner, H. (2019): Levels of Organic Life and the Human, New York: Fordham University Press.

Plessner, H. (2020): Laughing and Crying, Evanston: Northwestern University Press.

Renault, E. (2017): Social Suffering, London: Rowman & Littlefield International.

Roberts, W.C. (2023): Class in Theory, Class in Practice, in: Crisis & Critique, 10.1, 248–262.

Roth, G. (2015): Marxism in a Lost Century, Chicago: Brill.

Scott, R.L. (2022): The Limits of Recognition, in: Angelaki 27.6, 21–30.

Sennet, R. (1998): The Corrosion of Character, New York: W. W. Norton.

Sennett, R. and Cobb, J. (1972): The Hidden Injuries of Class, Cambridge: Cambridge University Press.

Theorie Communiste (2011): The Present Moment, in: Sic 1, 95–143.

Thompson, E.P. (1964): The Making of English Working Class, New York: Vintage.

Wright, E.O. (2023): Classes, London: Verso.

Dancing between Ideology Critique and Standpoint Epistemology

Rethinking Strategic Ignorance

Hilkje C. Hänel

Introduction

Recently, there has been a renewed interest in feminist standpoint theory. Under the headline of standpoint epistemology, these theories mash feminist standpoint theory with feminist critiques of science, such as Anderson's pragmatic approach to science (1995), Harding's strong objectivity (1993) or Longino's analysis of political values (1990). While feminist standpoint theory and feminist critiques of science share the idea of knowledge and knowers being situated—Harding (1993), for example, writes that we should start our investigations from the perspective of women—, they do not share more radical assumptions about knowledge and knowers. Feminist critiques of science are not in direct relation to Marxist materialist approaches nor to shared consciousness of oppressed social groups. Hence, it remains an open question how radical standpoint epistemology wants to be.

Here, I want to briefly discuss a difference between feminist standpoint theory and standpoint epistemology; claiming that standpoint epistemology has a less materialist approach to knowledge. I then focus on how critical knowledge can be achieved by marginalized and oppressed knowers despite—what I call—ideological ignorance and what could be described as false consciousness in Frankfurt School Critical Theory's toolbox. Next, I show that ideological ignorance is not the only way in which emancipatory knowledge is blocked. Instead, marginalized and oppressed knowers might decide—for many different reasons—to block access to emancipatory knowledge; a phenomenon that is sometimes labelled *strategic ignorance*. However, this seems to point to a tension between ideological ignorance and standpoint epistemology. If ideological ignorance can be tackled well by sharing knowledge and actualizing emancipatory and critical knowledge practices, it seems that any form of strategic ignorance is problematic for denying access to this deeply needed knowledge. In other words, if critical knowledge is not simply given but results from struggle and communal practices, then failing to engage in struggle or communal practices for strategic reasons blocks access to critical knowledge. It would then follow that strategic ignorance is problematic for our knowledge systems—a conclusion that I want to resist. Instead, I argue that strategic ignorance is a contextual phenomenon and that often practices of recognition and resistance should take priority over our responsibility for ameliorating the epistemic system. In fact, while considering the consequences for epistemic systems is important, this is often done on the backs of those who are oppressed; as concepts such as epistemic exploitation show.

Importantly, in this paper I am not interested in the way in which false consciousness or ideological distortions and ideological ignorance can be overcome and who is in a particularly good position to do so—even though those are important questions that I tackle elsewhere (Hänel 2025; see also Toole (forthcoming)). Instead, I am interested in (a) how false consciousness should be understood in a more complex way and in combination with cases of strategic ignorance and (b) the question of whether blocking access to critical knowledge—as can be the point in cases of strategic ignorance—is problematic.

1. From Standpoint Theory to Standpoint Epistemology

Marxist standpoint theory is an epistemological approach rooted in Karl Marx's materialist analysis of society. Roughly, it asserts that knowledge is socially situated, and that individuals' positions within the structures of capitalism fundamentally shape their access to and understanding of truth. The theory highlights the epistemic privilege of marginalized or oppressed groups, particularly the working class, arguing that these groups, by virtue of their material conditions, are better positioned to perceive and critique the contradictions of capitalist society. Standpoint theory includes the following key assumptions: First, that knowledge is socially situated; that is, all knowledge is in some form mediated by material social relations (for example, by relations pertaining to class, labor, and economic position). Second, that the oppressed (potentially) have epistemic privilege. The idea being that the lived experiences of oppressed groups offer a more complete and less distorted understanding of social totality; for example, the proletariat can (under certain conditions) unmask the ideological distortions of the ruling class (cf. Lukács 1923). Third, that knowledge can be gained to some extend through experience; the everyday life of oppressed people is not only a site of suffering but also a potential source of insight and critical awareness (cf. Smith 1987). And, finally, that standpoint theory's function lies in unmasking ideology and false consciousness insofar as dominant ideologies obscure the workings of capitalism and standpoint theory seeks to reveal these ideologies by grounding analysis in the material realities of those at the bottom of the hierarchy (cf. Marx & Engels 1976).[1]

Feminist standpoint theory was developed in response to Marxist and socialist feminist debates in the 1970s and early 1980s and draws significantly on Marx's historical materialism. The main aim was to understand relations of domination and exploitation not primarily with regard to the working class and their class consciousness, but with regard to gendered power relations. Feminist standpoint theory includes a rather diverse group of theories; for example Hartsock's historical materialist perspective of relations of domination as gendered (1983), Haraway's situated knowledges (1988), Collins' black feminist thought (2000), Smith's world sociology for women (1987, 1990), Hennessy's interpretation of standpoint theory as ideology critique (1993), MacKinnon's world making capacities (1982), and, more recently, Sandoval's third world feminist consciousness (2000); to name but a few.[2] What these

1 I am merely giving a brief overview here that should not be taken as conclusive. I will also not engage with critique of standpoint theory such as the potential problem of epistemic essentialism or relativism and the question of how standpoint theorists validate a "privileged" standpoint and how to account for intersectionality and heterogeneous experiences within oppressed groups; although I will touch on some of these below. See Collins (2000) as well as Rolin (2009) for a detailed discussion of feminist standpoint theory.

2 For a good overview, see Naples (2007). It should be noted that not all of the feminist standpoint theorists mentioned agree that feminist standpoint theory is in fact a specific methodology or that these theories share any significant overlap and some of them have been very vocal with regard to the limits of standpoint theory, for example Smith (1992). However, that these insights and theories share some general perspective can be seen in Harding's (2003) attempt to collect all influential essays on feminist standpoint theory. However, there are of course texts missing that have been written in different languages; German feminists writing in the Marxist tradition, for example, have been very involved in feminist standpoint theory (cf. Haug 1982, 1989; Scholz 2000).

attempts have in common is the acknowledgement that power dynamics influence knowledge in many important ways and the Marxist historical materialist view that "knowledge develops in a complicated and contradictory way from lived experiences and social historical context" (Naples 2007, 1). For example, Nancy Hartsock (1983) has adapted Marxist theory to argue for a specifically feminist standpoint that is grounded in women's (reproductive) labor and social roles. The sociologist Dorothy Smith (1987) has advanced a sociology rooted in the everyday experiences of women, emphasizing how dominant knowledge systems marginalize oppressed perspectives. And, Sandra Harding (1991) has expanded on standpoint theory by advocating "strong objectivity", arguing that starting research from the lives of the marginalized leads to more robust and objective knowledge.

The basic underlying idea of feminist standpoint theory as well as feminist critiques of science is the idea that knowers are not interchangeable. Rather, the social position of knowers is of epistemic relevance. Not only can differences in experiences lead to differences in perspective, but a differently located knower might not be able to have the same perspective even if they had the same experience. Feminist standpoint theory has expanded on this insight by arguing that differences in knowers are not random, but are socially structured, mediated by material social relations, such that experiences are differentiated along the lines of social location and social group membership. On the one hand, this implies that if we know the world and interact with it through our specific perspective, we only see a certain part of the world and are likely to miss other parts of it. Hence, our perspective not only shapes what we see and how we see it but also limits what we see. On the other hand, the importance of social location and social group membership is not that simple. It is neither the case that different experiences result in different knowledge nor that one's social location or one's social group membership results in some specific form of knowledge necessarily (cf. Intemann 2010, 783–4). Rather, as Gaile Pohlhaus summarizes, "the situations resulting from one's social positioning create 'common challenges' that constitute part of the knower's lived experience and so contribute to the context from which [they approach] the world" (2012, 716–7; cf. Alcoff 2000, 2006; Collins 2000). The argument is not that there is a direct link between having a certain experience or a certain social location or group membership and therefore necessarily obtaining some specific knowledge. Instead, what we develop are 'common challenges' or, in Heidi Grasswick's words, 'a perspective' (2018).

Recently, the insights of standpoint theory have seen a revival with both epistemology of ignorance and standpoint epistemology. Standpoint epistemology can best be understood as a critical theory of knowledge that asserts the epistemic significance of social location. Similar to what we have seen above, it challenges the Enlightenment ideal of a neutral, detached knower by positing that all knowledge is socially situated (Haraway 1988; Harding 1991). According to standpoint epistemologists, marginalized and oppressed groups, by virtue of their structural positions and/or their social identity within systems of domination, possess the potential for an epistemic advantage—particularly when it comes to understanding the power relations that shape social life (Hartsock 1983; Collins 2000).

Yet, analogue to the insights of standpoint theory, this epistemic privilege does not arise automatically from one's social identity but requires a process of critical reflection and and/or consciousness raising. A *standpoint*, in this sense, is not merely a perspective derived from experience, but a historically and socially achieved position from which critique of dominant ideologies and knowledge systems becomes possible (Harding 1991; Smith 1987). For example, Nancy Hartsock's (1983) foundational work develops a feminist standpoint grounded in women's labor and embodied experiences, drawing on Marxist notions of class consciousness. Similarly, Dorothy Smith (1987) emphasizes the epistemological significance of women's everyday experiences, arguing that mainstream sociology marginalizes these experiences through abstract, male-centered frameworks. Sandra Harding (1991) extends this logic by proposing the concept of "strong objectivity," claiming that beginning inquiry from the lives of the oppressed produces more rigorous and less distorted knowledge. Unlike traditional claims to objectivity, which often conceal the interests of dominant groups, strong objectivity embraces reflexivity and situates the knower within the structures of power they seek to understand. Patricia Hill Collins (2000), in her theory of Black feminist epistemology, further elaborates this approach by integrating the intersectional experiences of race, class, and gender, and emphasizing the communal and dialogical nature of knowledge production among marginalized groups.

At the core of standpoint epistemology, then, is the critique of the so-called "view from nowhere"—the idea that knowledge can be produced independently of social and political context (Haraway 1988). Instead, standpoint epistemologists argue that marginalized standpoints often yield more complete or less distorted insights into social reality because such groups must navigate both their own lived experiences and the dominant worldview. As

Pohlhaus (2012) writes, being in a position of vulnerability or dependence to those with more power implies that it is in the interest of the oppressed to know both their own experiences as well as the expectations of those in power. This dual consciousness, reminiscent of W. E. B. Du Bois's notion of "double consciousness," allows for a deeper understanding of social contradictions (Collins 2000).

While feminist standpoint theory and standpoint epistemology are closely related and often used interchangeably, they represent distinct, though overlapping, intellectual projects. Both challenge traditional notions of objectivity and neutrality in knowledge production, emphasizing the social situatedness of knowers. However, they differ in their scope, orientation, and theoretical commitments.

Feminist standpoint theory originates within Marxist feminist thought and is primarily concerned with constructing a political and epistemological framework that arises from women's material and social experiences. It posits that women, by virtue of their structural position within patriarchal or sexist societies, can develop a critical standpoint that reveals the gendered dimensions of knowledge and power. The theory is normative and emancipatory in orientation: it seeks not only to describe epistemic structures but to ground a feminist critique of science, society, and ideology (Hartsock 1983; Smith 1987). For example, Nancy Hartsock's seminal work on the feminist standpoint draws directly on Marxist theory, particularly on the notion that the working class can develop a "standpoint" that reflects the contradictions of capitalism more clearly than the bourgeoisie. Similarly, Hartsock argues that women, through their roles in reproduction and domestic labor, occupy a material position that offers epistemic insights into the structure of patriarchy (Hartsock 1983).

In contrast, standpoint epistemology is a broader philosophical and methodological approach that focuses on the epistemic implications of social positions and social identities. It is not limited to gender or feminism and is used across a range of critical traditions, including critical race theory, decolonial theory, and disability studies (Harding 1991; Collins 2000). Here, the focus is on how knowledge is produced, validated, and legitimized across different social locations. Moreover, standpoint epistemology is often concerned with developing meta-epistemological claims—that is, claims about the nature, limits, and structure of knowledge itself. For example, Harding (1991) and Collins (2000) argue that marginalized standpoints can reveal the partiality and bias of dominant knowledge systems calling for the systematic inclusion of multiple, especially marginalized, perspectives in the construction of knowledge. This marks a shift from the distinctly political project of feminist standpoint theory to a broader epistemological critique of the assumptions underlying Western knowledge production.

Feminist standpoint theory can be understood as a subtype of standpoint epistemology, with a more politically specific and normatively oriented agenda. Standpoint epistemology, by contrast, serves as a broader theoretical framework for analyzing how epistemic privilege and marginalization are structured by social position and social identity, and how these structures shape what counts as knowledge.

We can observe another conceptual shift from standpoint theory to standpoint epistemology that lies in the distinction between social position and social identity and that reflects deeper changes in both epistemological orientation and political strategy over time. Feminist standpoint theory, particularly in its early Marxist-influenced formulations, focuses on social position—understood in structural and material terms. Thinkers like Hartsock (1983) and Smith (1987) grounded their theories in women's economic, domestic, and reproductive labor. The emphasis was on the objective location of women within systems of patriarchy and capitalism, and how that location shaped epistemic access to social reality. In this framework, a *standpoint* was achieved through collective political struggle and consciousness-raising. In this sense, "social position" refers to a relational and structural category: one's place in the economic division of labor, family roles, institutional power structures, and so on. The focus is less on how individuals identify themselves and more on how systems of domination assign roles and locations; sometimes in contradiction with how individuals identify themselves.

By contrast, many contemporary versions of standpoint epistemology, particularly those influenced by poststructuralist or intersectional theories, place greater emphasis on social identity—as a subjectively meaningful, often self-ascribed category. These newer approaches draw from critical race theory, decolonial theory, LGBTQ+ studies, and intersectional feminism, and are more attuned to how individuals experience and articulate their own identities in relation to systems of power (Collins 2000; Alcoff 2006). Here, epistemic standpoints are increasingly linked to how individuals or groups identify themselves and how these identities shape lived experience, epistemic exclusion, and access to epistemic resources (Fricker 2007; Medina 2013). This emphasis on identity also tends to foreground narrative, testimonial, and affective experience as legitimate sources of knowledge—especially in contexts where dominant epistemologies have silenced or pathologized marginalized knowers.

The shift from social position to social identity reflects a broader transformation in critical epistemology—from a materialist, structural analysis of knowledge to a more cultural, discursive, and experiential one. While this shift has allowed standpoint epistemology to become more inclusive and responsive to intersectionality and identity-based exclusion, it also introduces tensions, in particular with regard to the achievement of standpoints. In the Marxist tradition, a standpoint had to be *achieved* through critical consciousness (cf. Lukács 1923). In contemporary formulations of standpoint epistemology, standpoints are often *ascribed* based on identity and experience; thus, raising questions of epistemic authority that I will come back to below.

Briana Toole's work, particularly in her 2020 paper, offers a reformulation of standpoint epistemology grounded in contemporary epistemological theory. Toole proposes a theory that centers social identity as an epistemically relevant factor—especially in explaining how credibility, authority, and access to knowledge are shaped by socially constructed and politically charged identity categories. Toole emphasizes that it is not simply one's objective social position—such as being part of a specific social class—that confers epistemic privilege or vulnerability, but rather how that position is mediated by one's lived social identity and its recognition within a given epistemic community. This allows her to draw on insights from theories of epistemic oppression (cf. Dotson 2011), where individuals are wronged in their capacity as knowers because of how they are relationally positioned in epistemic structures of inclusion and exclusion. Yet, while relying heavily on social identity in her account, Toole highlights the fact that standpoints are not necessarily given qua the knower's identity. She argues that standpoint epistemology consists of the following three theses:

(1) According to the situated knowledge thesis, whether an epistemic agent is in a position to know a certain proposition *p* depends on some non-epistemic social fact about that agent;
(2) according to the epistemic privilege thesis, some epistemic advantage can be drawn from positions of powerlessness; and
(3) according to the achievement thesis, knowledge accessible from a particular social location is not given, but must be struggled for.[3]

3 All three theses are taken from Toole 2019, 3.

Here, I want to take a closer look at the achievement thesis that "knowledge accessible from a particular social location is not given, but must be struggled for" (Toole 2019, 3; cf. Jaggar 1983, 383–4). In the feminist standpoint theory literature, the problem of achievement is well discussed; as Harding has argued, there is "no typical women's life" (1991, 10–1) such that all women have the same experiences; hence, standpoints of the lived experiences of women are necessarily heterogenous. And, women's lived experiences are not necessarily in line with feminist knowledge of women's lives (cf. Harding 1991, 123; Hennessy 1993, 14–5). In other words, whether or not a critical standpoint is achieved and whether or not such a standpoint captures robust knowledge of intersectional structures of oppression cannot be purely linked to a person's social group membership or social identity. Yet, taking a closer look at Frankfurt School Critical Theory can provide insight into how standpoints can be achieved and by whom.

In standpoint theory we can, roughly, identify two approaches to the problem of the standpoint of the working class. On the one hand, Frankfurt School Critical Theory broadly refers to the intellectual tradition that emerged from thinkers like Theodor Adorno, Max Horkheimer, and Herbert Marcuse. While it is often associated with critiques of culture and ideology, Frankfurt School Critical Theory in the context of the working class emphasizes the role of culture, ideology, and social norms in perpetuating capitalist structures. The focus here is on how capitalist societies maintain control through ideology and culture (media, education, etc.), rather than purely economic factors. This tradition highlights the cultural production of consent, arguing that the working class is oppressed not just through economic exploitation, but also through the ways in which culture and ideology are shaped by the ruling class. These critical Marxists argue that the working class's consciousness is shaped by alienation, lack of control over cultural production, and the way they internalize the dominant ideology; a critical standpoint has to be achieved by the working class itself. This standpoint is seen as crucial for critiquing dominant values and revealing the false consciousness that maintains capitalist power structures.

On the other hand, the Leninist approach to the standpoint of the working class has a more revolutionary focus, rooted in the writings of Lenin and his interpretation of Marxism. For Lenin, the working class's consciousness (or proletarian standpoint) was important, but it needed to be developed and advanced through external leadership. Lenin argued that the working class, in its immediate conditions, is often incapable of achieving revolutionary consciousness on its own due to its narrow economic interests and so-called 'trade union consciousness'. Lenin's approach focuses on the role of the vanguard party (led by a conscious, revolutionary elite) to guide and politically educate the working class towards revolutionary action. The standpoint of the working class is not just one of alienation or cultural critique, but a strategic position to mobilize political action that can overthrow the capitalist state. Hence, Leninists argue that the working class can only realize its revolutionary potential through strategic intervention and leadership—specifically, the party that acts in the interests of the working class. While critical theorists assumes that the working class has the potential for radical self-consciousness, though it is often stifled by ideology, according to the Leninist approach, the working class needs external guidance to move it towards revolutionary consciousness of its own power. It does not see the working class as inherently capable of achieving this consciousness on its own.

2. Between Achievement and Ideology: No Man's Land

To get a better understanding of the achievement thesis of feminist standpoint theory, I propose to look at it with the help of ideology critique; inspired by the Frankfurt School Critical Theory approach. The general idea being that we are all embedded in a pervasive ideological structure that shapes our lived experiences, our live projects, our social relations, and how we understand the world and our place in it. If this is the case, then what we do and how we understand what we do and who we are is at least to some extent distorted by the prevalent ideology; I have spelled this out in terms of sexist ideology (Hänel 2018, 2021, 2025), but it can and should be expanded to intersecting racist, white supremacist, ableist (etc.) ideologies. Understanding experiences of oppression happens within our embeddedness of these ideologies, significantly raising the bar for emancipatory or critical understanding. Since there is a wide range of theories and uses of the term 'ideology' and to avoid misunderstanding, let me give a brief overview of what I have in mind.

I understand ideology as a practical form of consciousness; that is as (a) a form of consciousness that (b) stands in relation to social practices. Ideology, according to this understanding, is not merely a distorted framework of the world, a problematic worldview, an illusion or delusion; it is not purely cognitive. Rather, it is a frame of intelligibility (or, in the words, of Marxist historical materialism, a system of beliefs) that is both *induced by social practices and has practical consequences*. In Rahel Jaeggi's words, "ideologies constitute our relation to the world and thus determine the horizons of our interpretation of the world, or the framework in which we understand both ourselves and the social conditions, and also the way we operate within these conditions" (2009, 64); hence, the frames of intelligibility motivate and rationalize our specific positions within society. According to Marx, the practical form of consciousness is determined by the economic conditions of a specific society. As social agents within this society, we adapt the economic relations that structure our lives and adopt a particular way of seeing the world and make sense of ourselves and our practices within that world according to virtues based on economic relations; hence the economic relations both make sense of our actions and our actions reproduce the very conditions and frames of intelligibility.

Understanding ideology as being comprised of the interplay of intelligibility frameworks and social practices directly relates it to questions of standpoint theory and epistemologies of ignorance. For example, according to Charles Mills, white ignorance—that is, the ideological frame of intelligibility that shapes, rationalizes and justifies white supremacy—causes white people to be ignorant of their own role as oppressors and their actions that (contribute to) oppress others. In this particular case, the frame of intelligibility naturalizes and, thus, legitimizes white supremacy, making it "easier" to participate in oppressive social practices. The general idea is simple: Depending on our place and the relations within that society, different rules, norms and expectations apply to us and are internalized by us, such that we reproduce the very frame of intelligibility that makes sense of our world in the first place. This is why, according to Barbara Fields, ideologies are a negotiated social terrain "whose map [we] keep alive in [our] minds by the collective, ritual repetition of the activities [we] must carry out in order to negotiate the terrain" (Fields 1990, 113).

It follows that if we are all in the grip of ideology, then questioning, critiquing, resisting the given frame of intelligibility—or merely seeing it for what it is—is far from easy. In fact, it requires to pause in our rituals and to turn away from what gives meaning to our lives; what makes us get out of bed in the morning. Let us consider an example to illustrate: The reason for why hardly one day goes by where I do not think that I did a bad job as a mother, because I try to juggle both two small kids and a full-time job and so neither job nor kids get the full attention they deserve, is largely to be found in the material conditions of the nuclear family and the norms and expectations that a sexist ideology constructs for what it means to be a good mother that govern my everyday life (even if I resist them): The ideal of a good mother is impossible to live up to; being attentive and calm while helping one child with their homework and happily engaging with the other one over some well-chosen toy that trains their imaginative skills, while also preparing a healthy afternoon snack for two hungry kids and a sugar-free cake for a school event the next day, while also doing the laundry and organizing a playdate for the weekend, while also re-scheduling the dentist appointment, and the list could go on. The expectations that a sexist and patriarchal society places on mothers are impossible expectations. Yet, the reason why I can turn to my girlfriends and complain about these impossible expectations or about the fact that being a mother is not my biggest fulfilment (despite this being what the sexist ideology makes it out to be) is because I am a *white* and *middle class* mother, which makes it less risky to complain about my life without anyone thinking that I do not love my kids or that social services need to be informed about a case of neglected kids and a struggling mother. A recent study shows that social services in Germany intervene more often and more drastically (for example, by taking kids away from their home) than in other cases when single mothers *ask* for financial help when there is no reported cause to do so (Hammer 2022). And this is despite the fact that single mothers are reported to be 4-times as likely to end in poverty (Lenze 2014). In the US, Black children and their parents are disproportionately reported to the child welfare system, often due to racist bias and discrimination (Boyd 2014, Harris 2021, Merritt 2021) with a gendered dimension playing an overlooked role (Breger 2012, Roberts 2014). The way in which the intersection of sexist, racist, and/or classist ideologies make intelligible what it means to be a good mother limits some mother's freedom even further, where already a complaint *about* the tasks associated with motherhood can be taken as an indicator that the wellbeing of her children is in danger.

Understanding the shortcomings of many mothers with respect to the ideals, norms, and expectations of motherhood not as an individual failure but as a structural problem—that is, as a distorted frame of intelligibility that brings it about that most of our maternal practices fall short of the ideal—is an achievement and not a given; even for mothers. Consciousness raising has developed as a strategy to collect knowledge and support women's political activism. Sharing their individual experiences of sexualized violence, domestic abuse and other forms of gendered oppression, women were able to see the commonalities of their experiences that then helped them to understand these experiences as structural rather than individual and to identify the social factors in place that shape these oppressive structures. This in turn helped to develop strategies against gendered oppression. The main idea, thus, is that knowledge about the particularities of women's lives can help to develop emancipatory strategies for social change. Catharine MacKinnon—as one proponent of early feminist standpoint theory who is directly influenced by Marxism—articulates an elaborate theory on how standpoints can be actualized through consciousness-raising groups (1991) and should, thus, not be taken for granted. Most recently, Toole (forthcoming) has developed an account of how consciousness-raising is a primary strategy to overcome dominant and distorted intelligibility frameworks.

Yet, Critical Theorists have long argued that people can be complicit in their own oppression and fail to access the specific knowledge related to their social position.[4] Wilhelm Reich famously stated "[w]hat has to be explained is not the fact that the man who is hungry steals or the fact that the man who is exploited strikes, but why the majority of those who are hungry *don't* steal and why the majority of those who are exploited *don't* strike" (1975, 53). Critical Theorists in the Marxist tradition answer by pointing to ideology, the product of a social system that engenders a form of consciousness that prevents its subjects from acting according to their own interests; as Adorno wrote, ideology is "necessary false consciousness" (1974 , 169).[5] When we speak of false consciousness, what we have in mind is *practical* false consciousness, not cognitive false consciousness or distortions of identity (cf. Rosen 1996, 72 for this distinction).[6] Practical false consciousness describes the way "in which we respond to and act within the world" (Rosen 1996, 72), for example, by being distorted in our beliefs, desires, interests, or will, in our values, ends or norms, or in our emotions. So, one Marxist explanation for why the hungry don't steal and the exploited don't strike is that they have a poor perception of their own interests. Their interests

4 Parts of this explanation are taken from my 2025 paper on standpoint theory and the relation between Critical Theory and non-ideal theory.

5 I should note that I do not take a stance on which theory of ideology is superior to others or whether ideology is nothing but a matter of false beliefs. For the purposes of this paper, I follow the idea that ideology is a practical form of consciousness (i.e., ideology is a system of belief—which are both true and false—that is induced by social practices and has practical consequences); as it was, for example, argued for by Rahel Jaeggi (2009). I will say more about this below.

6 This is not to say that an analysis of cognitive false consciousness or any other form cannot be useful in our theoretical endeavors. Yet, what I have in mind here is the way in which practical false consciousness relates both to our social practices and the ideological framework that justifies, legitimizes or naturalized the given social order; cognitive false consciousness, in comparison, remains on the level of cognition.

and "desires have been organized by a system which depends on their docility" (Meyerson 1991, 7) and have been shaped according to commercial values and consumer goods; here, their "immediate" interests are being taken for "real interests" (Marcuse 1964, xiii). The general idea being that we, as social actors embedded in ideological systems, fail to see what our real interests are and instead succumb to our immediate interests.

On the other hand, ideology is real in so far as it produces the actual social relations and practices in which we engage. This means conceiving of ideology as both real and unreal—or, quoting Adorno's famous remark, in ideologies "truth and untruth are always entwined" (1972, 465; translation in Jaeggi 2009, 66)—in so far as the distorted or false ideas of ideology leave an imprint on the social reality. This relates to the idea of "necessary false consciousness" because consciousness here is false in so far as it contains a false understanding of reality, yet, at the same time, it corresponds to reality by being socially induced.[7] Hence, being embedded within an ideology at least bears the risk of being ignorant to a critical or emancipatory understanding of the oppressive structures. A point also made by Charles Mills, when he writes about white ignorance: "producing the ironic outcome that whites will in general be unable to understand the world they themselves have made" (1997, 18). In fact, the way in which we understand the world and make our actions intelligible "needs to be scrutinized for its adequacy to the world, for how well it maps the reality it claims to be describing." (Mills 2007, 24) In other words, ideology maps a reality that does not in fact exist, while also being the terrain on which people are forced to live.

3. A Closer Look at Strategic Ignorance

So far, I have argued that ideology can lead to various forms of problematic ignorance such that those under its thrall fail to understand the world and their position in it adequately, while at the same time often actively (re-)producing

7 This was more recently described by Haslanger as so-called feedback-loops; that is, something becomes real because of what is attributed to it as when the identity of an agent is constructed by the social context providing certain concepts for their particular self-understanding, which are taken on board by the agent in question and evolve in line with their self-understanding (Haslanger 2012, 124; cf. Hacking 1988, 55).

the very conditions of oppression. This seems to imply that being critical of ideologies results in a better understanding of the world and its oppressive workings. Thus, anything that stops us from gaining a better understanding seems inherently problematic. In this section, however, I want to argue that the way in which false consciousness can hinder the achievement of epistemic privilege should be understood in a more complex way that allows for cases of strategic ignorance. The key idea being that ideological (or false) consciousness is not the only reason for why emancipatory or critical knowledge is not engaged with or shared. In this section, I have a look at cases of strategic ignorance, where a standpoint is achieved but not being shared with others and then turn to discuss the question of whether blocking access to critical knowledge—as is the point in cases of strategic ignorance—is in fact problematic because it restricts access to an adequate picture of reality.[8] In the next section, I argue that the question of whether holding back adequate knowledge in cases of strategic ignorance is problematic, is misguided insofar as practices of recognition should take priority over our responsibility for ameliorating the epistemic system.

Let us start by outlining specific social positions that can achieve epistemic privilege and consider one explanation of why this is the case. In his 2007 article on white ignorance, following W.E.B. Du Bois (1989), Mills presents three examples of double consciousness; the awareness of one's own experiences and identity and the understanding of how others see oneself or not see oneself and the expectations that follow. In *The Souls of Black Folk* (1989, 4), Du Bois writes:

> It dawned upon me with a certain suddenness that I was different from the others; or like, mayhap, in heart and life and longing, but shut out from their world by a vast veil.

A similar theme can be found in Ralph Ellison's *Invisible Man* (1995, 3), where he recounts being invisible:

> simply because [white] people refuse to see me … When they approach me they see only my surroundings, themselves, or figments of their imagination—indeed, everything and anything except me.

8 Raising the question of whether this could be problematic comes up in debates about the responsibility or duty of those oppressed to work against their oppression; see, for example, Boxill 2010.

And, as Mills details, Herman Melville's *Benito Cereno* (1986), where the protagonist boards a slave ship that was taken over by the Black slaves, who are holding the white crew hostage. For the protagonist, the very idea that Black slaves could accomplish such an act of insurrection is beyond the epistemic framework of white supremacy that he relies on, thus, he is trying desperately to find alternative explanations for the strange behavior on board of the ship; despite the many ways in which the actual evidence presents itself. In Mills' sharp words: "The white delusion of racial superiority insulates itself against refutation" (2007, 19).

The key insight then is, that asymmetrical relations of power produce incentives, on the one hand, to understand the world better for reasons of survival, and, on the other, to be ignorant of the social structures and unjust relations that make up the status quo of oppression. Mills uses the work of David Roediger (1998) to explain: Here, we have groups of Black and white people

> whose respective privilege and subordination tend to produce self-deception, bad faith, evasion, and misrepresentation, on the one hand, and more veridical perceptions, on the other hand. Thus [Roediger] cites James Weldon Johnson's remark "colored people of this country know and understand the white people better than the white people know and understand them" (5). Often for their very survival, blacks have been forced to become lay anthropologists, studying the strange culture, customs, and mind-set of the "white tribe" that has such frightening power over them (Mills 2007, 17).

Vulnerability, powerlessness or (problematic) dependence vis a vis others brings it about that one has to understand not merely one's own experience but also what the other can and will do and the assumptions, rules, and norms these actions are grounded in. In James Baldwin's words: "I have spent most of my life, after all, watching white people and outwitting them, so that I might survive" (1993, 217). Using the wording of "outwitting" suggests that while a critical standpoint was in fact achieved, the knowing subject choses (for reasons of survival) not to share the accumulated knowledge but instead to "play along" with the ignorance of the other. As Ella Surrey concludes: "We have always been the best actors in the world... We've always had to live two lives—one for them and one for ourselves" (Gwaltney 1993, 240). In fact, the writings of Black feminists provide not only insight into double consciousness, but also into the ways in which double consciousness can be tightly connected to strategic ignorance.

Both Patricia Hill Collins and bell hooks show how Black women are torn between being feminist and being anti-racist; in fact, they often have to give up on their feminist ideas and struggles to prioritize the more existential fight against racism.

There are good reasons for Black women to prioritize the fight against racism; and some of these reasons can be found in the shortcomings and failures of a prevalent white feminism (cf. Zack 2007 and 2005, Chapter 1). Mariana Ortega, for example, argues for what she calls "loving, knowing ignorance", the way in which white women exclude Black and brown women while at the same time wanting to include them. The loving feminist "knows (and wants to know) about" (2006, 61) Black and brown women; yet, the loving feminist is producing (often inaccurate) knowledge about Black and brown women to further their own end of being an inclusive and righteous feminist (cf. Ortega 2006, 58–61).[9] Ortega's theory of ignorance highlights the complexity of ignorance that is rooted in sexism, racism, and white supremacy while staying within the framework of ignorance as produced by the asymmetrical hierarchies of systems of power. In a similar spirit, Audre Lorde asks "[w]hat woman here is so enamored of her own oppression that she cannot see her heelprint upon another woman's face?" (1984, xx) and argues that white feminists are invested in ignoring the real differences between women (1984, 117–8; cf. Lugones 2003 and Spelman 1990). This is both a problem of injustice and a problem for our acquisition of knowledge. In fact, Gloria Anzaldúa writes that "[r]acism is especially rampant in places and people that produce knowledge" (1990, xix) and this is at least partly because of the intersection of sexism with racism and white supremacy, where dominantly situated women (read: white women) expect to have an epistemic advantage in so far as they are women, while also being "used to having conversations about racism and sexism in discursive spaces where [their] perceptions go unchallenged" (Bailey 2021, 59) and, thus, wrongly assume that their own position is "the 'default position'" (Spelman 1990, 13).

Yet, prioritizing the fight against racism over sexism is not without costs for Black women: Their experiences have always differed from the experiences of Black men to an extent that is hard to ignore or suppress (hooks 1990, 57–64). Gender is racialized. Collins shows how a Black gender ideology portrays Black

---

9 Interestingly, Ortega relates these observations to standpoint theory by asking how such loving, knowing ignorance can be possible despite claims of epistemic privilege of marginalized women; I will say more about this in the last two sections of the paper.

men as too weak and Black women as too strong; however, replacing the ideas of Black gender ideology with more emancipatory ideas of Black femininity and Black masculinity "will be difficult because elite groups have a vested interest in perpetuating ideologies of Black deviancy" (2004, 184–5). Furthermore, Collins shows that differences among African Americans cause different (racist and other) experiences; she writes "[e]ach Black person [...] had a common struggle, but the form it took differed greatly as well as our responses to it" (2004, 3). On the other hand, Collins makes explicit that there is a distinctive Black women's standpoint often expressed through "alternative ways of producing and validating knowledge" (2000, 252), what Collins calls 'subjugated knowledge' (251) based on lived experience and dialogue, which "has been devalued by dominant knowledge validation processes" (256) and, thus, "may not be claimed by many African-American women" (256). Both Collins and hooks were deeply troubled by and aware of the fact that the problem lies in how the intersection of racism and sexism dismisses the standpoint of Black women; not only of women and not only of Black people in general, but of *Black women*.

Let us relate these insights to accounts of strategic ignorance. Writing about Black social movements and, in particular, Black feminists' struggles, hooks starts her book *Ain't I a Woman?* with the following lines:

> Contemporary Black women could not join together to fight for women's rights because we did not see "womanhood" as an important aspect of our identity. Racist, sexist socialization had conditioned us to devalue our femaleness and to regard race as the only relevant label of identification. In other words, we were asked to deny a part of ourselves—and we did. Consequently, when the women's movement raised the issue of sexist oppression, we argued that sexism was insignificant in light of the harsher, more brutal reality of racism. (hooks 1982/1990, 1)

Here, hooks shows how a part of one's identity and, thus, a part of one's lived experiences can be *deliberately* devalued or suppressed. And in so doing, the Black women's standpoint is subsumed under the standpoint of Black experience. However, this is not due to ignorance. Rather, while the critical standpoint of what it means to be subjected to both sexism and racism (or white supremacy) has been achieved, the Black *woman*'s standpoint is not realized due to the deliberate decision that the knowledge underlying the critical standpoint of what it means to be a Black woman will not be shared with those outside of the social group.

As briefly mentioned before, this raises the following question: If it is the case, that to achieve a critical standpoint is an epistemic weapon against ideological and false consciousness because it unmasks the unjust social structures, then strategic ignorance for its power of omitting adequate knowledge is problematic. To explain: Accumulating knowledge about how the world is that maps reality adequately is necessary in dismantling unjust ideologies, hence, it seems to become a duty to feed critical standpoints into the epistemic system to allow more people to resist oppressive structures. According to this account then, being strategically ignorant means to be complicit in a system of oppression.[10] In fact, this assumption is implicitly made in the debate about epistemic injustices, when it is argued that testimonial and hermeneutical injustice are not merely bad and harmful to the person facing them, but to the wider epistemic community insofar as important knowledge is lost (cf. Fricker 2007).

4. Strategic Knowledge as a Form of Peer-Recognition

In the last section, I want to take a closer look at this problem and resist the assumption that strategic ignorance is problematic in the ways outlined above. My overall claim is that the question of whether holding back adequate knowledge in cases of strategic ignorance misguided insofar as practices of recognition should take priority over our responsibility for ameliorating the dominant epistemic system. First, let us consider two examples of strategic ignorance to gain a better understanding of what is at stake. In her article on strategic ignorance, Alison Bailey recounts research by Robin Kelly on Black working classes that shows that Southern Black laborers used the white assumptions according to which Black workers are lazy and incompetent to mask their calculated slowdowns of labor as laziness and their deliberate tool breaking as incompetence. Similarly, Ellison's *Invisible Man* gives an account of some of his teachers wearing chauffeur caps when driving through small surrounding towns to

10 I am leaving aside here the problem that many who have achieved a critical standpoint are not in positions of power which would allow them to feed their standpoints into the dominant epistemic system such that they became available to others as well as the very real and physical dangers that come with sharing critical knowledge. For now, I am interested in whether being strategically ignorant is problematic *assuming it is possible to share one's standpoint in some way or another.*

avoid trouble, well aware of the fact that whites assume that any good car is owned by a white person anyway (Bailey 2007, 88). These are cases of strategic ignorance that are grounded in a standpoint of racism and white supremacy and that makes use of said knowledge to both navigate the world unharmed (as in the case of masquerading as chauffeurs) or to resist the prevalent ideology without putting oneself in danger (as in the account of Southern Black workers).

Kristie Dotson develops the concept of *testimonial smothering* to describe cases in which a "speaker perceives [their] immediate audience as unwilling or unable to gain the appropriate uptake of proffered testimony" (2011, 244). In these situations, an oppressed or marginalized person truncates their "own testimony in order to insure that the testimony contains only content for which one's audience demonstrates testimonial competence" (ibid.). Dotson continues to outline the three conditions that can result in testimonial smothering: the content of the testimony is perceived as risky; the audience demonstrates or is known to be incompetent regarding said content; and, there is a form of pernicious ignorance at work (Dotson 2011, 244). Let us consider an example to illustrate. Black women sometimes engage in testimonial smothering about domestic or sexual violence within theor communities to avoid invoking problematic stereotypes of the "violent" black male in their white audience (Dotson 2011, 244–5; cf. Crenshaw 1991, 1256 and White 2001, 36). As Dotson shows, testimonial smothering only works if there is ignorance on the side of the audience and the knowledge about such ignorance on the side of the speaker; in other words, to be able to truncate one's testimony in the case above, one must be aware of the reality of domestic abuse as well as the ideological framework that includes the "violent" black male stereotype; while one's audience must be ignorant about these ideological stereotypes being stereotypes. Thus, this form of double consciousness grounds the decision to "hold back" knowledge.

Yet, there are two important aspects to these cases of strategic ignorance: First, the rationality behind "holding back" knowledge is the perceived danger that follows either from the content of the testimony or from the situation the person is engaged in. In other words, for *some* speaking the truth is riskier than for others. Second, strategic ignorance happens in contexts in which a person from a vulnerable position holds back knowledge from others who are positioned in such a way that they have power *over* the other person; and this holds even if said power is not actively used. Let us consider each aspect in turn.

First, as Kristie Dotson analyzes, situations in which testimony is given are shaped by dependency relations between speakers and audiences (Dotson 2011,

237; cf. Hornsby 1994, 134 on reciprocity in linguistic exchange). For such a situation to be successful, uptake on the side of the audience has to take place; that is, the audience has to understand the words that are uttered by the speaker *and* understand the meaning of these words as intended by the speaker. Dotson takes this to imply, and I follow this idea, that audiences have to make an effort to understand what the speaker is doing with their words (2011, 238). Hence, to successfully give testimony it is not merely important that a speaker receives credibility but also that the audience is "*willing and capable of hearing*" what the speaker tries to say. Receiving uptake under conditions of systemic ignorance can easily fail, especially when testimony is contradicting the dominant epistemic system that the audience has an interest in preserving. The more effort an audience is willing to take to preserve the (epistemic) status quo or the more embedded they are in the ideological form of consciousness, the harder and the more dangerous it will be for an oppressed speaker to try to receive uptake.

Second, many (if not most) situations of testimonial exchange happen under conditions where one person has more social power than the other. Holding back significant knowledge from those who are situated differently, can help navigate life better (and safer). Consider again the example from Ellison's *Invisible Man*: Sharing knowledge about how some Black men wear chauffeur caps when driving through small surrounding towns to avoid trouble at a specific time beyond the Black community could have meant to put the lives of these men in danger. Similarly, sharing accounts of domestic abuse beyond a trusted circle of other women can put a woman's life in danger. Yet, there are other situations, in which sharing the very same knowledge—for example to warn other women—can be equally important. In other words, whether strategic ignorance is in fact problematic for limiting the pool of accessible knowledge for the wider community or emancipatory for preserving the safety of others is a contextual question, in which the interplay between the content of what is said and the historically shaped power asymmetries between those in the conversation play crucial roles.

One way in which to understand this contextual framing of strategic ignorance is through the lens of recognition theory. As is well known from recognition theory, we—as social agents—rely on the recognition of others to become who we are, to form a positive relation to self, and to develop and maintain life plans. This is first and foremost a psychological insight; it tells us something about the conditions necessary for us to grow into full persons. Often, this claim is being put in relation to a theory of justice or a wider political project of the relation between state and individual, where recognition is assumed to be necessary for being equal and for being an individual member of a social group (cf. Honneth 1995, Taylor 1992). Yet, as recent critiques show, it is far from obvious why the psychological insight that recognition of others is a necessity to develop who we want to be has to be given within the context of (state) politics. As, for example, Coulthard (2014) has argued, recognition theorists have a tendency to think of recognition as a gift that is bestowed on the powerless by those in power. What is far more interesting, or so I have argued (cf. Hänel forthcoming), are claims of recognition *within* marginalized or oppressed communities. Here, we find important ways in which recognition and resistance or tightly connected.

The main idea is, that if we rely on the recognition of others to become agents, we also rely on the recognition of others to become epistemic agents; to be able to accumulate, use, and share knowledge. This aspect is particularly important for those, who are denied state recognition or who face conditions of oppression. To be capable of resisting a dangerous and hateful world, we need the recognition of our peers insofar as it lays the foundation for us to have a sense of who we actually are in this world; and not merely a distorted view of how those in positions of social power see and define us. We could say, double consciousness—as understanding who we are and how others see us—is only possible if we receive adequate recognition by our peers and communities and are, thus, able to form a positive relation to self, next to the distorted one that is imprinted on us by ideology. Furthermore, resistance requires accumulating ideology-critical knowledge and sharing knowledge; for example, the knowledge of harm suffered, the knowledge of how to engage in community building or to engage in acts of civil resistance. Elsewhere, I have described these forms of knowledge as *reflective knowledge* and *know-how knowledge* (cf. Hänel forthcoming). Reflective knowledge is what standpoint theory is interested in. Know-how knowledge, on the other hand, is simply what is needed for practices of resistance. Then, being strategically ignorant is grounded in reflective knowledge; as I have argued above, strategic ignorance relies on a critical understanding of one's position in the unjust social structure as well as the other's position, that comes with specific expectations, interests, and understandings of the world. And, being strategically ignorant can be important in certain contexts as it keeps know-how knowledge within certain communities and, thus, makes resistance possible in the first place.

5. A Very Short Conclusion

Before I end, I want to point out that this account of strategic ignorance fares well with the notion of epistemic exploitation brought forward by Nora Berenstain. Epistemic exploitation "occurs when privileged persons compel marginalized persons to educate them about the nature of their oppression" (2016, 569) and when the labor that goes into this education is "unrecognized, uncompensated, emotionally taxing [or] coerced labor" (ibid.). This can be the case, when the labor of teaching others is not compensated at all; in fact, often the labor that goes into such education is not even recognized *as labor*. It can also be the case, when the labor is recognized or compensated but the person positioned more powerfully then uses the knowledge to their own benefit. And, epistemic exploitation highlights the fact that the relevant kind of knowledge—what I have spelled out as *reflective* knowledge above—is often emotionally exhausting for those asked to educate others. Epistemic exploitation, according to Berenstain, plays a key part in maintaining the ideological system by reproducing problematic forms of ignorance insofar as those in powerful positions can use the shared knowledge "as fodder for skeptical responses and harvest them for content that can be developed into straw arguments and rejected out of hand" (2016, 587) while at the same time keeping those who share knowledge busy and, thus, distract them from epistemic resistance. Furthermore, those situated in powerful positions can mask epistemic exploitation as a "balanced, reasoned debate" (Berenstain 2016, 587), thus, making it harder still to dismantle the ideological system. Remember that I have started with the question of whether not sharing reflective knowledge can mean to by complicit in one's oppression insofar as knowledge about such oppression is necessary to unmask it. Berenstain provides another powerful argument for why *sharing* knowledge under conditions of asymmetrical power can mean to become complicit in the oppression as one delivers the insights necessary for others to uphold the status quo.

References

Adorno, Theodor (1974): "Ideologie," *Soziologische Exkurse*, Frankfurt a.M.: Europäische Verlagsanstalt.

Adorno, Theodor (1972): "Beitrag zur Ideologienlehre," in: *Soziologische Schriften I*, Gesammelte Schriften 8. Frankfurt a.M.: Suhrkamp.

Alcoff, Linda M. (2006): *Visible Identities: Race, Gender, and the Self*. New York: Oxford University Press.

Alcoff, Linda M. (2000): "On Judging Epistemic Credibility: Is Social Identity Relevant?" in: *Women of Color and Philosophy*, ed. Naomi Zack. Malden, Mass.: Blackwell Publishers.

Anderson, Elizabeth. 1995. "Feminist Epistemology: An Interpretation and a Defense," *Hypatia* 10(3): 50–84.

Anzaldúa, Gloria (1990): "Haciendo Caras: Una Entrada," in: *Making Face, Making Soul/Haciendo Caras: Creative and Critical Perspectives of Women of Color*, San Francisco: Aunt Lute Foundation Books.

Bailey, Alison (2021): *The Weight of Whiteness: A Feminist Engagement with Privilege, Race, and Ignorance*. Lanham: Rowman & Littlefield.

Bailey, Alison (2007): "Strategic Ignorance," in: *Race and Epistemologies of Ignorance*, ed. S. Sullivan & N. Tuana. Albany: State University of New York Press.

Baldwin, James (1993[1961]): *Nobody Knows My Name: More Notes of a Native Son*. New York: Vintage International.

Bambara, Toni Cade (1970): "On the Issue of Roles," in: *The Black Woman: An Anthology*, ed. T.C. Bambara. New York: Signet.

Berenstain, Nora (2016): "Epistemic Exploitation," *Ergo* 3(22): 569–90.

Breger, Melissa (2012): "The (In)visibility of Motherhood in Family Court Proceedings," *New York University Review of Law and Social Change* 36(4): 555–92.

Collins, Patricia H. (2004): *Black Sexual Politics: African Americans, Gender, and the New Racism*. New York & London: Routledge.

Collins, Patricia H. (2000): *Black Feminist Thought: Knowledge, Consciousness, and the Politics of Empowerment*, Second Edition. New York: Routledge.

Coulthard, Glen S. (2014): *Red Skin, White Masks: Rejecting the Colonial Politics of Recognition*. University of Minnesota Press.

Crenshaw, Kimberlé W. (1991): "Mapping the Margings: Intersectionality, Identitiy Politics, and Violence against Women of Color," *Stanford Law Review* 43(6): 1241–99.

Dotson, Kristie (2011): "Tracking Epistemic Violence, Tracking Practices of Silencing," *Hypatia* 26(3): 236–57.

Du Bois, W.E.B. (1989[1903]): *The Souls of Black Folk*. New York: Penguin Books.

Ellison, Ralph (1995[1952]): *Invisible Man*. New York: Vintage Books.

Fields, Barbara (1990): "Slavery, Race and Ideology in the United States of America," *New Left Review* 1(181): 95–118.

Fricker, Miranda (2007): *Epistemic Injustice: Power and the Ethics of Knowing*. Oxford: Oxford University Press.

Grasswick, Heidi (2018): "Feminist Social Epistemology," *The Stanford Encyclopedia of Philosophy*, ed. Edward N. Zalta, URL: https://plato.stanford.edu/archives/fall2018/entries/feminist-social-epistemology/.

Hacking, Ian (1988): "The Sociology of Knowledge about Child Abuse," *Nous* 22(1):53-63.

Hänel, Hilkje (forthcoming): "Homeplace: Recognition-giving Practices and Epistemic Agency in Communal Spaces," in: *The Philosophy of Recognition: Expanded Perspectives on a Fundamental Concept*, ed. M. Congdon & T. Khurana. Routledge.

Hänel, Hilkje (2025): "Non-Ideal Theory and Feminist Standpoints: The Problem of False Interests," *Journal of Social Philosophy* 56(1): 24–41.

Hänel, Hilkje (2021): "Who's to Blame? Hermeneutical Misfire, Forward-Looking Responsibility, and Collective Accountability," *Social Epistemology* 35(2): 173–84.

Hänel, Hilkje (2018): "What Is a Sexist Ideology? Or: Why Grace Didn't Leave," *Ergo* 5(34): 899–921.

Hammer, Wolfgang (2022): "Familienrecht in Deutschland: Eine Bestandsaufnahme," Bericht.

Haraway, Donna (1988): "Situated Knowledges: The Science Question in Feminism and Privilege of Partial Perspective," *Feminist Studies* 14(3): 575–99.

Harding, Sandra (2003): *The Feminist Standpoint Theory Reader: Intellectual and Political Controversies*. New York: Routledge.

Harding, Sandra (1993): "Rethinking Standpoint Epistemology: What is 'Strong Objectivity'?" in *Feminist Epistemologies*, ed. Linda Alcoff & Elizabeth Potter. New York & Oxford: Routledge.

Harding, Sandra (1991): *Whose Science? Whose Knowledge?: Thinking from Women's Lives*. Ithaca, NY: Cornell University Press.

Harris, Marian (2021): "Racial Bias as an Explanatory Factor for Racial Disproportionality and Disparities in Child Welfare," in: *Racial Racial Disproportionality and Disparities in the Child Welfare System*, ed. By A. Dettlaff. New York: Springer.

Hartsock, Nancy (1983): "The Feminist Standpoint: Developing the Ground for a Specifically Feminist Historical Materialism," in: *Discovering Reality: Feminist Perspectives on Epistemology, Metaphysics, Methodology, and the Philosophy of Science*, ed. Sandra Harding and Merrill Hintikka. Dordrecht: D. Reidel.

Haslanger, Sally (2012): *Resisting Reality: Social Construction and Social Critique.* Oxford: Oxford University Press.

Haug, Frigga (1989): "Das Verhältnis von Erfahrung und Theorie in feministischer Forschung," in *Videnskabshistorie og -teori. Kvindeforskningsseminar*, Kopenhagen.

Haug, Frigga (1982): "Erfahrung und Theorie," *Das Argument* 136.

Hennessy, Rosemary (1993): "Women's Lives/Feminist Knowledge: Feminist Standpoint as Ideology Critique," *Hypatia* 8(1): 14–34.

Honneth, Axel (1995): *The Struggle for Recognition: The Moral Grammar of Social Conflicts.* Cambridge, Mass.: MIT Press.

hooks, bell (1990): *Yearning: Race, Gender and Cultural Politics*. Boston: South End Press.

hooks, bell (1982/1990): *Ain't I a Woman: Black Women and Feminism.* Boston: South End Press.

Hornsby, Jennifer (1994): "Illocution and its Significance," in: *Foundations of Speech Act Theory*, ed. S.L. Tsohatzidis. New York: Routledge.

Intemann, Kristen (2010): "25 Years of Feminist Empiricism and Standpoint Theory: Where Are We Now?" *Hypatia* 25(4): 778–96.

Jaeggi, Rahel (2009): "Rethinking Ideology," in *New Waves in Political Philosophy*, ed. B. de Bruin and C.F. Zurn. New York: Palgrave Macmillan.

Jaggar, Alison (1983): *Feminist Politics and Human Nature*. Totowa: Rowan and Allenheld.

Lenze, Anna (2014): "Alleinerziehende unter Druck: Rechtliche Rahmenbedingungen, finanzielle Lage und Reformbedarf," Bericht, Bertelsmann Stiftung.

Longino, Helen (1990): *Science as Social Knowledge: Values and Objectivity in Scientific Inquiry*. Princeton, NJ: Princeton University Press.

Lorde, Audre (1984): *Sister Outsider.* Freedom (Calif.): Crossing Press.

Lugones, María (2003): *Pilgrimages/peregrinajes: Theorizing Coalition against Multiple Oppressions*. Lanham (Md.): Rowman & Littlefield.

Lukács, Georg (1923/1971): *History and Class Consciousness: Studies in Marxist Dialectics*, trans. Rodney Livingstone. London: Merlin Press.

MacKinnon, Catharine (1991): *Toward a Feminist Theory of the State.* Cambridge, Mass.: Harvard University Press.

MacKinnon, Catharine (1982): "Feminism, Marxism, Method, and the State: An Agenda for Theory," *Signs* 7(3): 515–44.

Marcuse, Herbert (1964): *One-Dimensional Man*. London: Routledge and Kegan Paul.

Marx, Karl and Friedrich Engels (1976): *The German Ideology, collected works*, vol. 5. New York: International Publishers.
Medina, José (2013): *The Epistemology of Resistance*. Oxford: Oxford University Press.
Melville, Herman (1986): "Benito Cereno," in: *Billy Budd, Sailor and Other Stories*. New York: Viking Pinguin.
Merritt, Darcey (2021): "Lived Experiences of Racism among Child Welfare-Involved Parents," *Race and Social Problems* 13: 63–72.
Meyerson, Denise (1991): *False Consciousness*. Oxford: Oxford University Press.
Mills, Charles (2007): "White Ignorance," in Shannon Sullivan & Nancy Tuana (eds.), *Race and Epistemologies of Ignorance*; 11–38. Albany, NY: State Univ of New York Press.
Mills, Charles (1997): *The Racial Contract*. Ithaca, NY: Cornell University Press.
Naples, Nancy (2007): "Feminist Standpoint Theory," in: *The Blackwell Encyclopedia of Sociology*, ed. George Ritzer. New York: John Wiley & Sons
Ortega, Mariana (2006): "Being Lovingly, Knowingly Ignorant: White Feminism and Women of Color," *Hypatia* 21(3): 56–74.
Pohlhaus, Gaile (2012): "Relational Knowing and Epistemic Injustice: Toward a Theory of *Willful Hermeneutical Ignorance*," *Hypatia* 27(4): 715–35.
Reich, William (1975): *The Mass Psychology of Fascism*. Trans. V.R. Carpagno. Harmondsworth: Penguin.
Roberts, Dorothy (2014): "Child Protection as Surveillance of African American Families," *Journal of Social Welfare and Family Law* 36(4): 426–37.
Roediger, David (1998): *Black on White: Black Writers on What It Means to Be White*. New York: Schocken Books.
Rolin, Kristina (2009): "Standpoint Theory as a Methodology for the Study of Power Relations," *Hypatia* 24(4): 218–26.
Rosen, Michael (1996): *On Voluntary Servitude: False Consciousness and the Theory of Ideology*. New York: Polity.
Sandoval, Chela (2000): *Methodology of the Oppressed*. St. Paul: University of Minnesota Press.
Scholz, Roswitha (2000): *Das Geschlecht des Kapitalismus: Feministische Theorie und die postmoderne Metamorphose des Patriarchats*. Horlemann.
Smith, Dorothy (1992): "Sociology from Women's Experience: A Reaffirmation," *Sociological Theory* 10(1): 88–98.
Smith, Dorothy (1990): *Conceptual Practices of Power.* Boston: Northeastern University Press.

Smith, Dorothy (1987): *The Everyday World as Problematic: A Feminist Sociology*. Toronto: University of Toronto Press.

Smith, Dorothy (1974): "Women's Perspective as a Radical Critique of Sociology," *Sociological Inquiry* 44: 7–13.

Spelman, Elizabeth (1990): *Inessential Woman: Problems of Exclusion in Feminist Thought*. Boston: Beacon Press.

Taylor, Charles (1992): "The Politics of Recognition," in *Multiculturalism*, ed. C. Taylor. Princeton: Princeton University Press.

Toole, Briana (forthcoming): "The Vanishing Self: Understanding Consciousness-Raising as an Epistemic Transformation."

Toole, Briana (2020): "Demarginalizing Standpoint Epistemology," *Episteme* (online first).

Toole, Briana (2019): "From Standpoint Epistemology to Epistemic Oppression," *Hypatia* 34 (4): 598–618.

White, E. Frances (2001): *Dark Continent of our Bodies: Black Feminism and the Politics of Respectability*. Philadelphia: Temple University Press.

Zack, Naomi (2007): "Can Third Wave Feminism be Inclusive? Intersectionality, its Problems, and New Directions," in: *The Blackwell Guide to Feminist Philosophy*, ed. Linda Alcoff and Eva Feder Kittay. New York: Blackwell.

Zack, Naomi (2005): *Inclusive Feminism: A Third Wave Theory of Women's Commonality*. Oxford: Rowman & Littlefield Publishers.

Domination, Recognition, and Discursive Control

Breno Santos

At the current stage of the scholarship concerned with the sociopolitical issues involved in how we share knowledge and make ourselves heard in knowledge-producing spaces, there has been increased attention directed at how institutional acknowledgment and integration can play a role in better shaping these spaces. The question of whether it is possible to think of adequate ways of sharing knowledge under the weight of socio-epistemic pathologies affecting our daily lives, such as epistemic injustices, structural ignorance, epistemic oppression, and exploitation, is central if we want to envision a path out of these pathologies.

One attempt to address this question is of note here and will be engaged with throughout the paper. It is Amandine Catala's (2015) discussion of hermeneutical domination, a type of epistemic injustice that affects the way individuals and groups make their contribution to institutionalized epistemic spaces, not because of knowledge or interpretive gaps on the part of the agent, but because of institutional power imbalance within these spaces. In Catala's view, hermeneutical domination is the injustice of dismissing a political minority group's epistemic contribution to the production of appropriate public discourse because of the lack of proper trust in deliberative spaces, making the minority group vulnerable to inadequate, but dominating and monopolized discourse on a particular social practice or experience. The way out of such a scenario, according to Catala, will involve a commitment to the basic requirements of deliberative democracy. She centers her account both on the potentially structural nature of epistemic injustices and on the institutional remedies to such pathologies.

In this paper, I want to engage with Catala's account via the recently highlighted connections between socio-epistemic problems and their potential origin in issues of recognition. Since the talk of epistemic injustices and domination, as we will see, revolves ostensibly around the failures to acknowledge the rational credentials and agencies of epistemic subjects, there's something to be said about the ways that these failures are, in fact, failures of recognition that entail a negation of relevant normative features of subjects, affecting how they are seen and how they see themselves, both epistemically and socially.

Building on the recent literature on this debate, I want to examine the connections between the talk of epistemic injustice as domination and the debate concerning the recognitive expectations we have as members of epistemic communities. In the first section, I want to explore Catala's formulation of hermeneutical domination and assess if it fits a recognitive model of the moral responsibilities we have towards other epistemic agents. I will argue that epistemic domination has its roots in failures of recognition. These failures, as we'll see, affect the agent's freedom and autonomy as a rational enquirer. Thus, the way out of them will need to involve non-domination in the form of recognition and discursive control, something that I will argue for in the second section of the paper. In the third and final section, I want to engage with a possible global version of hermeneutical domination that challenges the efficacy of local recognition and discursive control. Institutional remedies to this global type of epistemic injustice might not be able to account for the extent to which it affects our shared space of reasons with an overarching dominating power. To deal with the structural and institutional aspects of epistemic injustices and to secure non-domination and proper recognition, then, we'll have to go beyond the moderate, intra-institutional measures of deliberative democracy and propose radical measures to secure freedom and hermeneutical non-domination.

1. Hermeneutical domination as a failure of recognition

Following Miranda Fricker's (2007) efforts in conceptualizing some of the failures of our epistemic practices in terms of epistemic injustices, much has been said about how such failures or injustices have their origin in or are maintained through our basic institutional arrangements (Anderson 2012; Bohman 2012). How we organize our institutional lives, especially in educational and deliberative spaces, might directly impact the fairness of their epistemic outcomes. And might be more or less conducive to "indignities" (Congdon 2017: 243) that affect marginalized groups' and individuals' attempts to offer their epistemic contributions. Amandine Catala has identified one of such indignities that might affect people's ability to contribute to a society's pool of hermeneutical resources, particularly by offering socially situated understandings of dynamics of oppression, exploitation, injustice, violence, etc. What Catala calls hermeneutical domination is a type of epistemic injustice that excludes via the imbalance of political power in institutional settings.

With this concept, Catala adds to Fricker's basic notions of testimonial injustice and hermeneutical injustice and connects these injustices to the ways political power can present itself as epistemic power of domination that disenfranchises marginalized groups in their attempts to contribute to the understanding of a particular social practice or experience while leaving them vulnerable to the monopolized and imposed understandings of it. Hermeneutical domination, in her view, occurs through the combination of testimonial and hermeneutical injustices in an institutional setting. Someone suffers a testimonial injustice if one is unable to credibly testify on some particular matter due to some negative affect that the hearer or hearers might have on one's social identity (Fricker 2007: 1). A person who suffers this kind of injustice suffers a lack of acknowledgment of their competence or authority as an epistemic agent and, as a consequence of this, ends up with their epistemic standing compromised in their attempts to share knowledge in situations of unbalanced social and political power.

A potential and very likely outcome of such exclusion comes in the form of limited participation in the production of a society's general epistemic resources, including those that are central to the intelligibility of people's experience as members of socially marginalized groups. To the extent that some of their experiences remain unknown or not properly understood, this person is suffering what Fricker calls a hermeneutical injustice (Fricker 2007: 1). In a complex sociopolitical context, both injustices will most likely work in tandem, in a self-perpetuating vicious loop. Those who suffer testimonial injustices are likely to be the same ones to be marginalized from meaning-making practices, and once these practices are exclusionary, they'll tend to perpetuate the opacity of testimonial injustices by failing to come up with the epistemic resources to conceptualize them and the tools to address them properly.

Catala innovates by arguing for another level of injustice, one not considered by Fricker, that poses a great threat to some groups' attempts to have their hermeneutical contributions taken into account. The "differential levels of credibility and intelligibility [of] individuals based on their membership in different social groups" (Catala 2015: 424), which describe cases of testimonial injustices and hermeneutical injustices, respectively, prevent marginalized groups from having an equal footing in the way collective discourses on particular social practices are produced and shared, which make them subject to imposed discourses on said practices, in such a way that it is almost impossible for them to effectively contest to the wrong and wrongful picture of the practice. (Catala 2015: 429) Thus, by being forcibly – sometimes even by the force of law and violence (Santos 2021: 117) – subjected to monopolized understandings of these practices, this marginalized group is hermeneutically dominated by hegemonic sectors of society and their dominant epistemic frameworks.

It will be helpful to introduce a concrete example of how marginalized groups can be hermeneutically dominated, both to illustrate Catala's point and to allow us to see further theoretical upshots in terms of the debate on the notions of recognition and social visibility, which will be addressed in section 2. Consider the following example[1]. A popular politician ascends to power, partially because of his strong anti-LGBTQ+ rhetoric. Both during his campaign for office and his term, he attempts to paint LGBTQ activists and researchers on topics related to LGBTQ rights as morally deviant and as posing some danger to young children and society as a whole, suggesting that they want to impose what he sees as deviant and promiscuous education upon children – he calls this a "gender ideology". Once in power, not only does this politician continue with the same rhetoric, but also supports and proposes laws that make it illegal to teach anything that challenges deep-seated religious commitments on gender and sexuality in this society. He even suggests that his supporters should interrupt classes, record teachers, and, if necessary, use violence to "save our children" from those he sees as degenerates. In such a scenario, defending LGBTQ+ people's and serious gender scholars' perspectives on these issues becomes more and more dangerous, with the threat of physical and institutional violence looming over whoever attempts to question traditional values in that society.

I think that this example expresses a high-stakes case of hermeneutical domination. But, and this will be important later on, although possibly widespread and a product of overarching ideology, it is related specifically to a particular discourse and targeted to a select group of people. It relates to discourse on gender, and it affects specifically people who have either first-hand or scholarly authority (or who are allies) on the relevant issues. The identity of the speaker, the content of their testimony, and, sometimes, their expressive style make them vulnerable to the three types of injustices discussed so far. They can suffer testimonial injustice when testifying on gender issues, by

1 I first introduced this example in Santos 2021: 116.

having their credibility deflated to below the minimum threshold. They can also suffer hermeneutical injustice because of the systematic rejection they might suffer on the testimonial level, which can preclude their contribution to the production and revision of that society's epistemic resources.

And, more importantly, they are victims of hermeneutical domination to the extent that even when they have valuable contributions to make to the collective hermeneutical pool on the debate on gender, their contribution is blocked, sometimes by institutional measures, through the threat of violence or via their insufficient participation on knowledge and meaning-producing spaces, given the tendency to political ostracism and segregation imposed by the hegemonic powers that be. So, this case seems to fit well with Catala's account of hermeneutical domination[2].

All the cases that Catala presents in her paper seem to revolve around similar aspects of this case. They involve the exercise of political power to limit marginalized groups' and individuals' movements in social-epistemic spaces. In most of her original cases, this is expressed in deliberative impasses concerning a particular practice or social experience. In the example above, we can cash this limitation in the imposed restriction on the institutional practices of LGBTQ+ people and/or competent researchers on LGBTQ+ issues of making sense of, and educating others on, the material reality and experiences of these marginalized groups. They are limited in their discursive power under the weight of dominant institutional discourse on gender and sexuality, and under the threat of institutional sanctions and potential violence.

As it was gestured before, the recent discussion on how recognition theories can help us understand how someone can be harmed in their rational agency might be of great help to us in our attempt to account for the moral harms inflicted upon those who suffer epistemic injustices and for the normative demands to correct those. I think that both the epistemological expansion of Axel Honneth's theory of recognition and new Hegelian accounts of the ethical character of the "knower" will also help us identify what is wrong with the case of hermeneutical domination we've just discussed.

A current theme in this recent literature concerns the importance of social mediation when we interact with other people in our daily epistemic practices. (Giladi 2018, 2022; Congdon, 2017, 2018) Knowledge practices, in these views, are directly conformed by social mediation in the sense that it is through this mediation, and its social and political contents, that individuals and groups

2 See Santos 2021 for a longer treatment of this idea.

are afforded authority, cognitive environments are established, cognitive resources are distributed, and epistemic power is consolidated. Epistemic injustices, then, would involve one or more instances of inadequate social mediation in the space of reasons.[3]. Thus, some persons' or groups' very statuses as rational participants in epistemic practices are products of the workings of this social mediation.

This is a Hegelian assumption that drives some of the work in accommodating the character and normativity of our epistemic lives in terms of relations of recognition. Since "Hegelian agency starts from the insight that we are always already part of a social community, and thus subject to a range of discursive, social and cultural norms and practices" (Schuppert 2013: 897), these accounts draw from Hegel and other recognition theorists in their attempt to diagnose "epistemic injustice as a social pathology" (Giladi 2017: 1) and to offer a view of the normativity involved in acknowledging one as having relevant statuses concerning knowledge-production and sharing in a community.

Matthew Congdon (2017, 2018) offers us an interesting incursion into this relation. He analyses epistemic injustices in terms of the violation of basic recognitive expectations that should guide this social mediation[4]. Congdon is interested in uncovering the "deeper indignity at work" (Congdon 2017: 243) when someone suffers an epistemic injustice. This indignity, he argues, comes in the form of the denial of social validation demanded by the very concept of a "knower". To call someone a "knower" is to position this person in the space of reasons in a qualified way, since such act "(i) ascribes agency to its target and (ii) situates that agency within social practices of giving and asking for reasons." (Congdon 2018: 4) Thus, to be a knower is to be expected to perform knowledge-related tasks rationally, responsibly and reciprocally.

The normativity of 'knower' implies its ethical dimension in two ways. On the one hand, to *regard oneself* as a knower is to see oneself as possessing "certain normative powers and responsibilities in the context of practices of justification and warrant" (Congdon 2018: 6), it is to regard oneself as fit to perform the relevant knowledge-related tasks and to value oneself for being able to do so. It is, in Congdon's account, to have "a distinctive sort of positive self-relation" (Congdon 2018: 6), one that merits special ethical treatment reserved for those who are active members of epistemic communities. In addition to that, to *regard someone else* as a knower is to perceive this person as both being in a

3 More on that idea of the "space of reasons" in what follows.

4 Not his term.

position to perform the relevant knowledge-related tasks and being criticized for their performance, but also to see them as having this positive self-relation.

Alternatively, we could say that a knower is a species of rational agent in the sense that the knower is capable of reason-giving and of acknowledging others in their normative features, especially in their features as rational agents or reason-giving actors (Schuppert 2013: 898).

Now, given this normative picture of the "knower", how can we cash out our responsibilities towards knowers? Congdon addresses this by offering a retooling of traditional Hegelian recognition theory. In this theory, at least in its more popular version, that advanced by Axel Honneth (1996, 2007), to recognize someone is both to acknowledge this person as having relevant normative status and also to act accordingly upon such acknowledgment in a way that socially validates the person's status. (Congdon 2018: 8) Here, the possibility of having a positive self-relation is directly dependent on whether one is part of a "developmental process of socialization whereby others confer recognition of their worth and social standing through affirmative expressive gestures and acts" (Honneth 2001: 115; Congdon 2018: 8). So, in this picture, rational agency requires mutual recognition (Schuppert 2013: 896). To regard oneself as a knower involves having normative expectations of proper ethical treatment, in the sense that one's epistemic authority should be perceived, and such perception should lead to proper action that acknowledges such status. The absence of actions of this sort, or the presence of actions that violate this status, is equivalent to the denial of proper recognition and can lead to moral and epistemic injury. Paul Giladi puts it similarly when he talks about intersubjective recognition as a process "identical to the journey of self-realization as a social and rational agent". (Giladi 2017: 4) The mutuality of recognition, then, is part of the Hegelian picture being sought by these accounts. It intends to express the aspects of social mediation, the socio-relational mechanisms that give rise to agency, rationality, autonomy, etc. (Schuppert 2013: 898), not only from an external but also from an internal perspective.

Thus, to regard one as a knower is to confer on this person a particular status as rational enquirer (Giladi 2017: 5), one that implies a set of normative expectations, be that in terms of evidence-gathering, testimonial competence, reason-giving, and perhaps even knowledge-communication clarity and accuracy.

Epistemic injustices, then, are ways of denying that an agent or a group can be expected to attend to these expectations; it is, as Giladi puts it, a way of robbing them of the status of rational enquires, of robbing their "rightful place as someone moving in the space of reasons" (Giladi 2017: 6). They do that when they violate one or more of what, according to contemporary recognition theorists, are the three basic recognitive expectations that someone, in their attempt to move within the shared space of reasons, has towards one's interlocutors. To be a knower is to have one's needs or particularities, one's particular competence, and standing as bearing general knowledge-related abilities acknowledged. When one doesn't have these features acknowledged, according to Congdon, one suffers instances of epistemic neglect, epistemic disesteem, and epistemic disrespect, respectively. (Congdon 2018: 15)

Love (standing here as the opposite of neglect), respect, and esteem are then the three basic forms of social validation that should be conferred through the social mediation of our epistemic endeavors. First, as an epistemic agent, one requires integration in the epistemic community of knowers in the sense of having guaranteed one's access to basic standards of education and initiation in the shared space of reasons. Second, one requires a basic presumption of goodwill on equal footing to other agents involved in daily knowledge practices – a basic form of respect for a person as a person, independently of one's track record on a particular topic or a particular knowledge-related task. Third, one requires proper acknowledgment of one's achievements and particular abilities for navigating the epistemic commons, through proper and just evaluation of one's competence and sincerity in particular topics or tasks. And, by the same token, the reciprocal self-relation that comes from having these needs attended is, in the epistemic realm, self-love, self-respect, and self-esteem. (Congdon 2018: 11)

Epistemic injustices, then, involve failures of affording to an individual or a group one or more of these three basic forms of recognition. Let us go back to our main example and see whether this normative framework can be put to use to account for hermeneutical domination. In the example, discourse on gender is being precluded by the powers that be that, via institutional measures, including the force of law and the threat of violence, in such a way that blocks contestation when such attempts come from particularly located subjects, who are conspiratorially seen as deviant, promiscuous and as a threat to society's "proper functioning". Those concerned with the correct account, or at least committed to a critical assessment, of gender norms in defiance of the patriarchal, sexist, homophobic, transphobic status quo, are seen as non-trustworthy on the topic at hand, having their identity institutionally associated with deviancy and epistemic risk. Because of that, they are blocked from relevant testimonial settings, where both contestation and explanatory resource-building usually take place. This lack of trust, I suggest, expresses itself as both a form of epistemic disrespect and epistemic disesteem.

On the one hand, there's a deflationary attitude towards these individuals' humanity, including its epistemic features, since the stereotype of the dangerous deviant subject lowers one's standing as an equal in this particular society. On the other hand, by having their authoritative input denied, those with lived experiences or theoretical baggage on the questions at hand are not met with the expected esteem for their relevant epistemic expertise. One could also add that, particularly to those directly affected by gender and sexual violence, the failure to take their (sometimes life-or-death) contributions to these questions is also a type of epistemic neglect that obscures their needs and particularities in the shared space of reasons.[5]. Thus, hermeneutical domination seems to fit in the recognitive model advanced by Congdon and other recognition theorists. So, beyond the possible accumulated harms of the other types of epistemic injustices, hermeneutical domination also involves epistemic neglect, epistemic disrespect, and epistemic disesteem. Together, these mistreatments amount to the denial of a basic status as rational inquirer and, as a secondary effect, they compromise these agents very capacity of self-conception, not necessarily in terms of hermeneutical lacunae – as it can be the case when one suffers a hermeneutical injustice (Giladi 2018: 12) –, but in terms of their sense of belonging to the shared space of reasons. One can end up unmotivated to partake in the collective knowledge tasks, once one is seldom in equal standing as those who maintain hegemonic perspectives in these spaces. However, in what follows, I want to suggest that hermeneutical domination not only involves failures of recognition but is a case of domination *specifically because of that.*

2. Freedom as democratic recognition

As we saw, Catala argues that hermeneutical domination, like the other basic types of epistemic injustices, is the product of inequalities of epistemic power. (Catala 2015: 436) The imbalance of epistemic power affects the way people's trustworthiness is assessed and, as a consequence of this, their ability

---

5 It's possible to think also of neglect as a long-term effect of domination since new generations will be precluded from proper epistemic participation because of hermeneutical domination.

to equally participate in "the production of the appropriate public discourse and policy regarding the social practice [is] at stake". (Catala 2015: 425) All the cases Catala discusses involve an escalating process of credibility and intelligibility deficits that give rise to the unequal hermeneutical conditions present in hermeneutical domination. Both in her cases and mine, there's an interpretative impasse affecting the way a particular social practice or a set of practices is understood. In both cases, marginalized groups are subjected to the monopoly of public discourse access and to actual corrupted public discourse that is imposed by those in hegemonic positions in society.

The disenfranchisement of marginalized groups makes them vulnerable to dominant epistemic practices that are imposed upon them in a way that fails to track these groups' relevant interests and blocks the effective possibility of contestation. (Catala 2015: 430) Drawing from the debate on domination outside of the epistemic realm (e.g., Bohman 2012), as Catala also does, we can understand domination as involving a "severe imbalance of power between two agents that is left unchecked, creating the capacity of arbitrary interference". (Catala 2015: 430) Non-domination, then, would naturally involve the effective dismantling of the possibility of arbitrary interference. In the case of hermeneutical non-domination, it would involve a correction of such imbalance and dismantling of monopoly over society's hermeneutical resources, to allow for control and contestation power in the discursive arena. (Catala 2015: 430)

Looking at the idea of domination more broadly, in the way it relates to freedom in a broad sense, the notion of freedom as non-domination can be spelled out in different ways, but one that points to some of the concerns of those who are worried about the effects of the hermeneutical domination is that of *freedom as discursive control* that can be found in some recent works on the republican tradition (Bohman op. cit.). To be free, in this perspective, is to have one's rational agency protected against arbitrary interference. (Schuppert 2013: 896) Freedom here is the ability to be in discursive control, it is to "have the ability to discourse and… have access to discourse" (Schuppert 2013: 896), and, as Fabian Schuppert points out, "In order to be in discursive control, people have to stand in discourse-friendly relationships to each other, as only in such relationships can agents exercise their agency freely." (Schuppert 2013: 896) According to Philip Pettit's account of freedom, one that's nicely summarized by Schuppert (2013), to be free is to be able to participate in "discourse as a critical, reflective reason-responsive person", but to do that in a "discourse-friendly" environment, with the possibility of having one's reason-giving abilities socially recognized (Schuppert 2013: 897).

Now, the demand for recognition is central in this account, since both from the republican perspective and for the new-Hegelian recognition theorists, our full agency is only realized in social reason-giving practices. For the latter, particularly, for these practices to be successful, they need to involve mutual recognition in the terms already discussed so far. So, to be free is to move qualitatively in the shared space of reasons, where one's abilities, needs, and social standing are – or can be – properly acknowledged in self-relation and in relation with others (what Schuppert calls the 'inner' and 'outer' conditions of freedom as discourse control, respectively [Schuppert 2013: 898]). In this picture, "discourse-friendly relationships can be redescribed as relationships of recognition, in which persons are socially recognized as reason-responsive and reflecting discourse participants". (Schuppert 2013: 896) To be free, in this sense, is to be recognized as a rational agent in a way that allows for one's autonomy to be present in the form of responsible endorsement. For Schuppert,

> reciprocal intersubjective recognition was defined as a complex social-relational state in which two important conditions are fulfilled: first, every agent recognizes the other members of society as competent judges and legitimate sources of claims and reasons, while enjoying reciprocally this very same recognition as a legitimate reason-giver; second, every agent sees himself or herself as a legitimate source of reasons, meaning that an agent can stand by his or her judgment and take responsibility for it. (Schuppert 2013: 899)

Therefore, hermeneutical domination would be a way of impinging on general freedom by compromising someone's agency and ability of discursive control and responsibility. By not being able to significantly weigh in on discourse aimed at explaining particular social practices, one is dominated not only because one is coerced into imposed understandings of a given practice, but because, as a consequence of a set of failures of recognition, one is not free, in a broader sense. One doesn't stand in a position to move freely in the basic space of agency exercise, the discursive space, the shared space of reasons, because one is not respected, esteemed, and loved. So, for our hermeneutical domination case, we saw before, the policies and propaganda advanced to criminalized debate on gender inflicts upon the discourse environment in two ways: it disrespects those involved in questioning hegemonic discourses, making the environment not friendly for the proper exercise of discursive control; and, as a product of that, it puts them under the dominating power of the hegemonic understanding that the hegemonic forces advance, and thus effectively dominating those who are marginalized.

From the above, it seems that to secure freedom in the space of reasons, one needs first to secure proper recognition. Catala's proposed adoption of three basic deliberative democracy criteria to counter the effects of hermeneutical domination seems then to be called for here. For her, to secure epistemic trust, we need to observe three substantial criteria of democratic deliberation that aim to control, limit, and distribute hermeneutical power: equality, legitimacy, and accountability. (Catala 2015: 424)

First, treating marginalized groups and individuals as equal in their epistemic status is a way of correcting the "credibility imbalance" they face in the discursive space and of distributing epistemic power. Fricker argues for something along these lines when she argues for the *virtue of testimonial justice*, which is, in her account, central to undoing testimonial injustices[6]. Putting this in the language of recognitive expectations, we could suggest virtually the same thing, but in terms of basic epistemic respect, the egalitarian portion of the epistemic recognition fits nicely with the first criterion Catala proposes.

Second, acknowledging the legitimacy of expert discourse is a way of limiting epistemic power only to those who have the relevant credentials on the matter at hand. In the case of marginalized groups, their members have first-hand experience that might "[result] in social expertise or epistemic privilege. Thus, a commitment to legitimacy in the context of deliberative democracy requires recognizing minorities' special credibility or epistemic privilege, which allows contestation." (Catala 2015: 437) The locus classicus of such an idea, one highlighted by both Catala and Fricker, is feminist standpoint theory and its view of the potential special credibility of marginalized groups in accounting for their oppressive relations. Again, in recognitive terms, epistemic esteem as an acknowledgment of expertise qualifies the agent to produce specialized discourse.

Third, by securing an accountability condition in the space of reasons, we control epistemic power through transparent reason-giving, especially when the reasons come from those who are marginalized. The idea here, for Catala, is that their skin-in-the-game-ness makes them reliably accountable to standards of reason-giving practices in rational discursive and deliberative spaces.

6 See Anderson (2012) for an institutional version of this idea.

Thus, in the recognition theory parlance, it is a way of avoiding epistemic neglect in the form of obliviousness to groups' and individuals' needs and particularities.

Once we have secured these three criteria, a deliberative space becomes a space of institutional recognition, a democratic deliberative space where non-domination is instituted. The result is that political minorities will have their epistemic abilities respected, their special expertise recognized, and their particularities accounted for in the debate, making them no more vulnerable to coercive and arbitrary interference of hegemonic hermeneutical power. This can be the case, for example, when a progressive force takes power with the intent of dismantling the oppressive mechanisms put forward by bigoted forces. Once new democratic measures are put to work, measures that involve these criteria, the political minority regains the ability to contest the remaining bigoted practices and understandings. Not only does the minority cease to be dominated, but also it ceases to be misrecognized, or not to be recognized at all. The minority is respected and esteemed, but it is also epistemically considered (loved), which in turn allows for its members to build or reinforce their processes of self-relation, especially in terms of the development of intellectual autonomy as the aforementioned responsible endorsement that is sought in one of the reciprocal ends of recognition. It seems, then, that the three deliberative democracy criteria, equality, legitimacy, and accountability, proposed by Catala as a way of resolving the deliberative impasse caused by hermeneutical domination, match some of the accounts put forward by recognition theories to explain the conditions not only for the proper ethical relationship but also for freedom in the space of reasons. In the remainder of this paper, I'll present a potential problem for this deliberative picture of freedom as recognition.

3. Global hermeneutical domination

It's important to locate the epistemic injustices debate within the broader debate on structural oppression and institutional constraints on one's ability to properly exercise one's agency in the face of these oppressions. As we'll shortly see, especially when talking about hermeneutical domination, the scope and reach of the epistemic pathology might influence the scope and reach of the proposed solutions. Although Catala's deliberative democracy framework offers the institutional route in the direction of discursive freedom as recognition – after all, securing the criteria of equality, legitimacy, and accountability seems to be an important step towards the constitution of a healthy and free discursive environment –, I want to suggest that there's a different type of hermeneutical domination shaping discursive environments in a way that makes it harder for us to transpose the same solution and to envision *fully free* discursive practices.

Consider, for example, a case of what I will call *global hermeneutical domination,* as in the case of Charles Mills' (1997) account of the Racial Contract. It would take us more time and space than we have available to present a detailed explanation of Mills' account, so let me explore just the main features, which I think are enough for us to grasp the intended analysis in terms of domination. In his 1997 book and subsequent work (2007, 2015), Mills puts forward a critical political philosophy centered on the uncovering of the racial (and racist) roots of traditional contractualist political theories. His critique, however, aims not only at these theories but also at racialized outlooks producing and being produced by such theoretical views. In his account, white supremacy has been the prevalent political ideology shaping the way our societies have organized themselves and their self-images in the past centuries. As the prevalent political and ideological force, white supremacy shapes the way we conceptualize basic ethical and epistemological notions, such as "human" and "knower" for example, the way we perceive ourselves and each other, i.e. as less or more fitting to these basic notions, especially in the assessment of other people's epistemic authority, and also the way some information becomes part of a society's collective memory, i.e. as central explanatory tools to account for the shape of human relations throughout history. The theoretical outcome of white supremacy's normative power (or white normativity) is what Mills calls a Racial Contract. As his 1997 most famous passage asserts,

> Thus in effect, on matters related to race, the Racial Contract prescribes for its signatories an inverted epistemology, an epistemology of ignorance, a particular pattern of localized and global cognitive dysfunctions (which are psychologically and socially functional), producing the ironic outcome that whites will in general be unable to understand the world they themselves have made. (Mills 1997: 18)

Thus, one of the practical results of the Racial Contract is widespread and actionable ignorance. Corrupted forms of consciousness and understanding, which ignore the true character of marginalized, non-white peoples' material existence, on the one hand; and impose white-interested outlooks to account for the way the world is or should be, on the other. Those on the losing end of the Racial Contract, I want to propose, are hermeneutically dominated. Given that, according to Mills, the influence of the Racial Contract is widespread across racialized societies, there'll be a tendency in those societies and their institutions to see the world not in its true colors, especially when it comes to seeing the relations of oppression, exploitation, and marginalization that shape most of the social relations taking place in them. But, more importantly, white normativity will determine who can speak and what one can speak about, which will in turn affect the proper participation of racialized individuals in our daily epistemic practices. The "ideological socialization" (Mills 1997: 81) of the Racial Contract will, in effect, monopolize both the distribution of epistemic authorities and the criteria for what can count as evidence in the space of reasons.

Objecting to these corrupted outlooks will most likely come with the burden of sometimes having to swim upstream the institutional current on particular discourses, as in the case discussed above, what I'm now calling *localized hermeneutical domination*. But also – and more importantly – for its global form comes with the challenge of presenting alternative understandings to a potentially infinite set of practices, concepts, perspectives, etc., all affected by the way white normativity has been, in the last centuries, shaping our cognitive tools, our meaning-making strategies and shared spaces of reasons, the organizational schema of political institutions and so on. The hermeneutical domination here presents itself in the form of evidentially resistant outlooks *imposed on the whole of society*, but also in the sedimented nature of the corrupted cognitive patterns that we tend to inherit and that affect the way testimonial and hermeneutical contributions are made and assessed in a racialized society. If one wants to contest or object to the main ideological accounts of the ways race has shaped our society and its institutions, including its deliberative spaces, one will face an uphill challenge. White normativity is the dominant explanatory force and, together with its associated theoretical tools, it constantly affects people's very ability to contest it in a robust way (i.e., in a way that threatens the status quo).

So, we have two illustrations of hermeneutical domination, one local and one global. Both seem to involve the monopoly and imposition of particular discourses and the near impossibility of contestation. What separates them, I suggest, is their scope of influence. While the "gender ideology" discourse functions by blocking particular inputs on a specific matter, in specific discursive spaces, the Racial Contract shapes the way we think, remember, conceptualize, and perceive, and also the way we set up our institutions, not only in their composition but also in their principles, goals and basic functions. It is the structural nature of this global hermeneutical domination that, as we will see, makes it difficult to resolve through the same deliberative mechanisms that Catala advocates for the local cases (which are the cases she seems to have in mind in her work). This is the case because this global type of domination, I will argue, infringes upon the very conditions for the fulfillment of the recognitive expectations involved in her democratic remedy.

Under global hermeneutical domination, it seems that it is only possible to secure trust via the criteria of equality, legitimacy, and accountability, and their recognitive correlates of respect, esteem, and love, in a *partial* way. This is because global hermeneutical domination is constantly threatening the undoing of progressive measures taken in well-intentioned democratic spaces. *Political integration*, of the type Catala proposes, in a racist society might be a way of securing immediate acknowledgment and participation, but it will not secure them in all close-by possible worlds. In scenarios of global hermeneutical domination, there's no undoing of the mechanisms of domination by integration, since the threat of mistreatment will always loom over all.[7].

Given that white supremacist ideology presents itself as white normativity and shapes our collective epistemic lives, be that in terms of theory-building or public discourse, it is natural to think that those marginally located will have a hard time contesting and opposing basic racialized commitments present in these societies, especially given the tendency of, in a racist society, those who are racialized being seen as inapt to advance some knowledge claims. I think this is correct. As in the case of local domination, those marginalized and authoritative (from experience or research) individuals who try to advance alternative interpretations of the social phenomena will be met with institutional blockades put in place to deflate any threat to the highly advantageous (to those in hegemonic positions) status quo. Thus, these individuals will most likely have their standing as apt contributors to the hermeneutical pool doubted, their expertise denied, and their skin-in-the-game-ness neglected, composing the same picture of ethical mistreatment as seen in the above case of failure of recognition.

---

7 We could compare this to the distinction Marx makes in *On the Jewish Question* (1975) between political emancipation (through access to liberal rights) and human emancipation (not secured by mere liberal rights).

However, the global case involves a further problem. Given that white supremacy and white normativity are prevalent forces in shaping racialized societies, in most of their interactional and institutional aspects, there's a case to be made that, although racially marginalized individuals are on the receiving end of most of the harms of the epistemic pathologies thus produced, the inability of contestation is shared across society more or less independently of the identity or the marginalized status of the those who attempt to contest. The ideological force of white normativity subjects *everyone* to its corrupted outlook and its arbitrary interference in the space of reasons. Again, this is not to say that everyone is harmed in the same way. But the near impossibility of contestation is a central feature of the ways the shared space of reasons is constituted. Such a feature makes hermeneutical domination a *built-in mechanism* that directs a great number of our social and institutional interactions.

One still could argue that the same is the case with local hermeneutical domination. After all, *anyone* who tries to oppose the sexist, homophobic, transphobic, and queerphobic policies will be met with disdain, at best, and violence, at worst. But notice that, although this might be the case, it is only contingent on the example. This society might elect a progressive candidate instead of a bigoted one next time, and the institutional measures to block gender discourse contestation might be overturned. This won't mean the end of bigoted gender discourse, since it fundamentally involves deep-seated gender ideology, but it can mean the end of the local hermeneutical domination through discursive integration. It's important to highlight that the domination in the local case is the product of actively imposed discourse and policies targeting a particular group identity or the content of this group's discourse. In the global case, however, hermeneutical domination is not tied to specific policies, but it constitutes the very discursive fabric of the racist society, regardless of those who are in power.

James Bohman (2012) argues for something along these lines when he defends that epistemic injustice as domination should not be focused on an identity prejudice explanation for the systematic character of the phenomenon. For him, an alternative explanation should point to the idea that the systematicity found in most cases of epistemic injustices is the product of domination. As Nancy Hartsock puts it, and Bohman quotes it, "the dominated live in a world structured by others for their purposes—purposes that at the very least are not their own and that are in various degrees inimical to our development and even existence". (Hartsock 1998: 241; Bohman 2012: 178) Thus, in his view, domination lies at the very basis of the injustice and affects the subject's standing as knowers by making them vulnerable to epistemic harm. Domination is the "systemic social hand" that influences our prejudiced attitudes towards marginalized groups and makes it such that these attitudes track these groups in their daily lives, making epistemic injustice one of the features of this exclusionary system (Bohman 2012: 177). Thus, epistemic injustice is one of the expressions of domination, not the opposite. There's no identity power, his argument goes, without a prior power of domination, be that as patriarchal domination, the continuation of slavery through segregation – and, I would put it – the Racial Contract and white normativity; there's no identity power without the political power of denying people "basic statuses and powers". (Bohman 2012: 183)

Although Bohman means to generalize this assessment to all cases of epistemic injustices, we don't need to commit to this stronger thesis. But what I have tried to show is that there's at least one type of hermeneutical domination that fits this structural pattern. As Bohman puts it, "deeper inequalities have sustained exclusions from the broad public sphere in which hermeneutical participation might have taken place" (Bohman 2012: 184). In this sense, global hermeneutical domination is the structural ideological manifestation of political domination as a whole. And, because of that, it needs to be analyzed taking this complexity into account.

Therefore, global hermeneutical domination can be seen as the ideological side of the political constraints put upon people by the power of arbitrarily interfering in their basic statuses, including their epistemic status. The Racial Contract not only directs the way we think about race but also does it by giving shape to the social institutions we take part in daily, even if these institutions, from time to time, work to integrate those marginalized from public discourse. Even in these cases, the institutional discursive environment is never discourse-friendly. They might look friendly on the superficial level, but once the threat of arbitrary interference is always there, after all, we have yet to see the end of racist social relations, *only* via integration, one never ceases to be dominated. By the same token, one is never in the position of being truly respected, esteemed, and loved. There might be fleeting acknowledgment, but never full recognition. Full recognition is only possible in a true discursive-friendly environment, one where freedom as discursive control means exactly freedom as non-domination.

From this picture, it seems that the moral import that recognition theorists propose runs deeper than one might first think. Since the challenges to contestation are widespread, failures of recognition are not necessarily indexed to marginalized identities. Marginalized identities will be systematically harmed by white normativity, but the primary wrong of global hermeneutical domination as an *epistemic* injustice lies in the arbitrary control of *the society's ability to contest* and the self- and other-relational aspects that come with it. In a society where white supremacy is a hegemonic force, the pool of hermeneutical resources is dominated as a feature of the society. Its epistemic spaces and its political institutions will all be established under disrespectful and neglectful assumptions towards first the marginalized groups, but also anyone who tries to question the racial basis of this society. Global hermeneutical domination will give rise to epistemic neglect, disrespect, and disesteem, but these are not classic recognition failures. They are structural failures of recognition and should be met as such.

Given the above, there seems to be a need to improve upon Catala's proposal of non-domination as just deliberative democracy. I don't have the space to argue for this here, but it seems that we need institutions that not only protect us from individual, local cases of domination, but we need institutions that are born out of commitments of non-domination. They might be the fruit of marginalized-based constitutions. They might be reshaped by processes of popular consultation. Or they might be born and shaped via a revolutionary movement. In any case, truly free institutions are still not among us. Until then, any discursive dynamic within them is vulnerable to hermeneutical domination.

4. Concluding remarks

In this paper, my aim was twofold. First, I wanted to further the connection that recognition theorists identified between epistemic injustices and failures of recognition. To do that, I explored Amandine Catala's notion of hermeneutical domination, a combination of two basic forms of epistemic injustice that function as ways of blocking proper contestation in relevant testimonial settings. I argued that hermeneutical domination involves a deflationary attitude directed to marginalized people's humanity and the denial of authoritative contributions of their lived experiences or expertise, resulting as a result epistemic disrespect and lack of esteem towards the target group or individual. On top of that, such blockage tends to amount to a type of epistemic neglect of the needs and particularities of marginalized groups. These individuals end up with their basic statuses as rational inquirers denied, and their capacity for self-conception as rational agents compromised.

I also argued that hermeneutical domination not only involved failures of recognition, as specified above but that it is an instance of domination precisely because of that. If we think of domination as a way of interfering with the general freedom of an epistemic agent, by blocking their ability of discursive control in a deliberative space, we have to locate this lack of freedom on the lack of proper recognition, which would guarantee the conditions of epistemic trust via the criteria for democratic deliberation and its correlate recognitive measures: treating them as equals in their epistemic statuses by epistemically respecting them, acknowledging the legitimacy of their expert discourse by epistemically esteeming their capacities as epistemic agents, and securing accountability in the space of reasons avoiding the epistemic neglect of their needs and particularities as people who have a deep interest, because the risk it represents to them otherwise, of the correct account of their marginalization. Once secured, the three deliberative conditions proposed by Catala and the three recognitive conditions proposed by the theories of recognition, together, promote not only a space of institutional recognition but also freedom as a feature of deliberative practice.

Finally, although not exhaustively, I argued for the need to improve upon Catala's account of non-domination as just deliberative democracy by taking into consideration the threat of a global type of hermeneutical domination as arbitrary interference and control of the whole society's ability to proper contestation regarding deep-seated ideological outlooks that shape our institutions and deliberative spaces as a whole. Once the failures of recognition are structural and not related to a particular topic and a particular deliberative setting, we might need structural remedies to them that don't seem to be readily available in the accounts of hermeneutical domination and recognition taken separately.

References

Anderson, E. (2012): Epistemic justice as a virtue of social institutions. *Social Epistemology*, 26(2), 163–173. https://doi.org/10.1080/02691728.2011.652211

Bohman, J. (2012): Domination, epistemic injustice and republican epistemology. *Social Epistemology*, 26(2), 175–187. https://doi.org/10.1080/02691728.2011.652217

Catala, A. (2015): Democracy, trust, and epistemic justice. *The Monist*, *98*(4), 424–440.

Fricker, M. (2007): *Epistemic injustice: Power and the ethics of knowing*. Oxford University Press.

Congdon, M. (2017): 'What's Wrong with Epistemic Injustice?', in I. J. Kidd, J. Medina, and G. Pohlhaus, Jr. (eds.) *Routledge Handbook to Epistemic Injustice*. New York: Routledge.

Congdon, M. (2018): "Knower" as an ethical concept: From epistemic agency to mutual recognition. *Feminist Philosophy Quarterly*, 4(4).

Giladi, P. (2018): Epistemic injustice: a role for recognition?. *Philosophy & Social Criticism*, 44(2), 141–158.

Giladi, P., & McMillan, N. (Eds.). (2022): *Epistemic injustice and the philosophy of recognition*. Taylor & Francis.

Hartsock, N. C. M. (1998): *The feminist standpoint revisited and other essays*. Boulder, CO: Westview Press.

Honneth, A. (1996): *The struggle for recognition: The moral grammar of social conflicts*. Cambridge, MA: MIT Press.

Honneth, A. (2001): Invisibility: On the epistemology of recognition. *Aristotelian Society Supplementary*, 75(1), 111–126.

Honneth, A. (2007): Recognition as ideology. *Recognition and power: Axel Honneth and the tradition of critical social theory*, 323–347.

Marx, K. (1975): *On the Jewish Question*. In K. Marx & F. Engels, *Collected Works* (Vol. 3). New York: International Publishers.

Mills, C. W. (1997): *The racial contract*. Ithaca, NY: Cornell University Press.

Mills, C. W. (2007): White ignorance. In S. Sullivan & N. Tuana (Eds.), *Race and epistemologies of ignorance* (pp. 11–38). Albany: State University of New York Press.

Mills, C. W. (2015): Global white ignorance. In M. Sullivan & A. Qureshi (Eds.), *The philosophy of race* (pp. 109–130). New York: Columbia University Press.

Pettit, P. (2001): A Theory of Freedom. Oxford: Oxford University Press.

Santos, B. R. (2021): Echo chambers, ignorance and domination. *Social epistemology*, *35*(2), 109–119.

Schuppert, F. (2013): *Freedom, recognition and non-domination: a republican theory of (global) justice* (Vol. 12). Science & Business Media. Dordrecht: Springer

Love's Labour Lost

Cases of Testimonial Betrayal

Clara Berlich

Introduction

Imagine a woman, Angela, who has lunch with two of her colleagues, Ernie and Bert. At lunch, Angela tells them that their boss sexually harasses her. (Let's assume he repeatedly makes unwarranted comments about Angela's looks and seems to "accidentally" touch her a lot.) Both, Ernie and Bert, are positive that Angela is being oversensitive, under a lot of stress, and that their boss is a very nice guy who can do no wrong.

The exemplary case outlined above falls under the category of what Miranda Fricker (2007) terms testimonial injustice.[1] Testimonial injustice typically occurs when a hearer ascribes lesser credibility to a speaker's testimony due to an unfair identity prejudice, as when Bert and Ernie fail to believe Angela.[2] Jeremy Wanderer (2017) notes that there are cases of testi-

1 Note that Fricker uses a similar exemplary case evolving around the testimony of sexual harassment to explore *not* testimonial injustice but to introduce another form of epistemic injustice, *viz.* hermeneutical injustice. However, the scope of my enquiry is limited, and I shall bracket all concerns regarding hermeneutical injustice here.

2 In the framework of this paper, I assume that cases of testimonial injustice *typically* involve a hearer who ascribes lesser credibility to the speaker's testimony and that an identity prejudice is at the core of the flawed epistemic process. Note that testimonial injustice may also include cases of credibility excess (for example, see Davis 2016) and that some authors argue that we can give an account of testimonial injustice without an appeal to prejudice (for example, see Perrine 2023). My discussion is thus limited: Whether my findings are also applicable to cases of testimonial injustice characterized by an unfair ascription of too much credibility or to an account of testimonial injustice as not involving any identity prejudice on part of the hearer, remains an open question that I cannot address here.

monial injustice, which "critically involve thick relations of intimacy between acquaintances" (cf. Wanderer, 2017: 36). Wanderer's observation is as shocking as it may seem trivial. If testimonial injustice is as pertinent as Fricker takes it to be, it makes sense to assume that it spreads to the realm of intimate relationships. Yet, it is bad enough if stereotyping and prejudice lead hearers to unfairly reject a stranger's testimony. If the same thing happens between friends, lovers, or family members, where people know and trust and perhaps even love each other, things seem to get truly ugly – and also truly complex. [3] Assume that Angela and Bert go way back, they have been part of the same close-knit group of friends for six years. Angela and Ernie on the other hand are simply colleagues, they have known each other for a couple of months and have never met off hours. Intuitively, Bert's unjust treatment of Angela's testimony strikes one as even worse than Ernie's failure to believe her. The following inquiry is devoted to account for this intuition.[4] More specifically, I want to propose a conception of cases of testimonial injustice within the context of intimate relationships as *testimonial betrayal*.

In the context of differentiating between varieties of testimonial injustice, Wanderer (2017) already provides a brief sketch of such a conception. Here, I set out to discuss and develop Wanderer's ideas further. To this end, I utilize Margalit's (2017) work on betrayal to gain a better understanding of what exactly constitutes the 'betrayal-part' in testimonial betrayal. Further, I set out to explore some basic tenets from recognition theory to highlight the distinct moral weight that identity prejudices bear within the sphere of intimate relationships. I argue that the crucial difference between pure cases of testimonial injustice and cases within the context of intimate relationships (and the explanation for the latter being in some relevant way worse) is first and foremost this: the latter undermines the meaning of a significant relationship, which simply does not obtain within the framework of the former.

During the course of this essay, I hope to shed light on how what we (epistemically) owe each other in intimate relationships and what we (epistemically)

---

3 In a similar fashion, Kristie Dotson (2011) provides an analysis of the relations of dependence speakers have on audiences, assuming (as I also do in this paper) that such a relation of dependency within intimate relationships is marked not primarily by vulnerability but by trust.

4 Note that testimonial injustice does not require fully fleshed disbelief, but only an unfairly low attribution of credibility. In the following, I mostly focus on cases of disbelief for reasons of simplicity.

owe each other in a broader social context are interdependent questions. This essay thus also intends to make a small contribution to feminist debates evolving around the question how social pathologies and political power asymmetries operate at and function through the sphere of human intimacy, thereby rendering the distinction between public, political spheres of our lives on the one hand and private, apolitical ones on the other, untenable.

Before I start, let me concede that the exemplary case that sets the tone for the remainder of my paper is a very simple one, involving an identity prejudice held against a woman qua her being a woman.[5] As has been pointed out by others, the injustice involved in testimonial injustice is often and critically of intersectional nature, featuring more than the one kind of marginalization, which I explicitly address in the following. I hope that the findings of my inquiry are applicable to other, more complicated cases than merely to the comparatively simple one I discuss. Yet this hope is a humble one, as I am aware of how these aforementioned simplifications necessarily limit the horizon of the essay.

1. A Special Case of Testimonial Injustice

When Ernie does not believe his colleague, Angela, that she is being sexually harassed and the reason for his misjudgment originates in a prejudice against her social identity as a woman in a world dominated by (white) men, this seems utterly unfair. When, for the same reason, Bert does not believe his friend that she is being sexually harassed, this does not only seem utterly unfair, but it in some way *exceeds* the category of unfair. How can the perceived difference between Bert's and Ernie's respective treatment of Angela's testimony be spelled out? Broadly speaking, there are two ways to reply to this question. The first would be that Bert's case is worse because it features *more* of the *same* kind of badness as involved in Ernie's case. The second would be that Bert's case bears an *additional distinct* kind of badness that does not feature in Ernie's case. I believe that neither of these replies are strictly speaking false. However, I want

5 Additionally, note how both hearers in my exemplary case are (identifying as) male. Overall, I believe that the case of testimonial injustice I am trying to make in the following also holds for a non-male hearer rejecting an intimate's testimony based on an unfair identity prejudice held against the speaker. Yet there may be subtle but crucial differences in this regard, which the scope of this enquiry does not permit to explore.

to argue that Bert's case (and other cases in the context of intimate relationships) necessarily exhibits an additional kind of badness that is distinct from the badness of other instantiations of testimonial injustice.

In this section however, let me first consider the option that cases of testimonial injustice in the context of intimate relationships feature more of the same kind of badness involved in testimonial injustice per se. Intimate relationships put aside for the moment, how precisely should we understand the harm or wrong involved? Since Fricker's influential book, much debate has evolved around the question of what makes instances of testimonial injustice bad and in how far certain understandings of what is so bad about testimonial injustice may enable or complicate the quest for its remedies (see e.g. Crawford 2021; Pettigrew forthcoming). I will not delve deeply into this debate here, although I hope that a closer look at the badness of a certain subclass of cases of testimonial injustice can, by a somewhat questionable logic of inference, teach us something about the bigger picture. I will come back to that point later. For now, let me first start with a few quick notes on what Fricker herself has to say about the badness of testimonial injustice. I then try to see if we can account for the particular badness of Bert's case within the framework that Fricker herself provides. After I demonstrate that Bert's case does not necessarily involve a higher degree of epistemic illegitimacy, I move beyond Fricker's formula and consider whether a lack of empathy is what can explain our intuition about the badness of Bert's case. I argue that it is not, but that it does point us into the right direction. Finally, I suggest that recent attempts to build conceptual bridges between recognition theory and feminist epistemology can supplement and enhance Fricker's account. That is, I argue that recognition theory does not only help us to get a fuller idea of the badness of testimonial injustice per se, but critically helps to see what makes testimonial injustice in the realm of intimate relationships so particularly harrowing.

1.1 Intimate Relationships, Testimony and Prejudice

If you do not believe me, when I say that I came up with a definition of knowledge that rules out Gettier cases because you know that I am drunk, this does not constitute an instance of testimonial injustice. If you do not believe me because I am a woman and you are convinced that women are bad at philosophy, it does. Importantly, even if I am drunk and completely off the rail whilst speaking, but your main reason for rejecting my testimony is not this but me being a woman, this still constitutes a form of testimonial injustice. Note that your prejudice, at least on Fricker's original account, probably does not operate on the level of conscious belief, but rather stems from a broader source of social imagination influencing your beliefs and resistance towards adopting beliefs on a subconscious level.[6] Fricker targets a *categorical connection* between the social practice of testimony and a form of injustice rather than an instantiation of injustice associated with some acts of testimony (cf. Wanderer 2017: 28). As your prejudice holds not only in my case, but presumably for all women you know, you are *systematically* (and unjustly) excluding women from the common enterprise of gaining and sharing knowledge about philosophy. Yet, Fricker's concept of testimonial injustice tracks not only the unfairness of this systematic exclusion itself, but also highlights the "symbolic weight" this kind of exclusion carries. If a speaker is systematically excluded from the social practice that "originally generates the very idea of a knower", this renders the speaker to be less than a full epistemic subject (Fricker, 2007: 17). Following Fricker, what this comes down to is the objectification of speakers, because they are treated as mere sources of information instead of receiving the right kind of acknowledgement and respect in their capacity as a knower (cf. Fricker, 2007: 132).

There sure are many open questions concerning the badness of testimonial injustice and the precise nature of the harm or wrong inflicted in cases thereof. Are we dealing with forms of moral or epistemic badness here, or both? Why is it bad for someone to be objectified in their capacity as a speaker? However, I think the assumption that testimonial injustice is *somehow* bad is well motivated by now, and we are also already in a position to grant that testimonial injustice in the context of intimate relationships is *somehow* even worse. One way to do this would be to focus not on the harm inflicted on the speaker but rather on the degree of epistemic illegitimacy or irresponsibility involved in the hearer's rejection of testimony. In general, any "hearer's obligation is obvious: she must match the level of credibility she attributes to her interlocutor to the evidence that he is offering the truth" (Fricker, 2007: 12). The difference between a stranger and somebody I know is first and foremost exactly this: a

6 Several authors point out that testimonial injustice may also operate consciously or even be prompted by a fully conscious intention on part of the hearer (for example, see Hänel 2024). In the framework of this paper, I simply assume that there are many cases of testimonial injustice, *some* of which are caused by an identity prejudice of which the holder is less than fully conscious. Exceptional but plausible cases set aside, I think that what happens if testimonial injustice occurs within the intimate context of a friendship or any loving relationship, may typically rather involve prejudices operating at a less than fully conscious level.

stranger is a person I do not know. To know someone includes to have at least some share of information about that person. This starts with a person's name and more trivial facts about them, and, in cases of intimate relationships, expands to all kinds of facts about that person. Hence, if I am prejudiced against an intimate friend, the prejudice itself might denote a graver epistemic and/or moral wrong than in other cases of testimonial injustice. Returning to the case of Bert and Angela, who have been friends for six years, we might wonder how it is even possible for Bert to maintain a prejudice when he had six long years to experience Angela as a person, that is, six long years of collecting all forms of facts and evidence about the qualities Angela has; both as an individual person and as an epistemic agent. Thus, we may believe that Bert is in a decidedly better position to assess the testimony at hand in this case than when assessing the testimony of a stranger. But when Angela gives testimony of being the victim of sexual harassment, Bert's prejudice towards women outweighs all of the evidence he should have about Angela being a credible speaker. Here, it seems like the prejudice on whose grounds Bert ascribes an unduly low degree of credibility is measured against everything Bert (should) know about his friend. In the end, it is an empirical question how much counterevidence there really is with regard to the purportedly low credibility of a friend's testimony. After all, it is very likely that in each and every instantiation of testimonial injustice, there would have been a sufficiently high amount of counterevidence to the prejudice operating when the hearer rejects a speaker's testimony. Further, if I have a friend who is a woman, knowing her and thus having some sort of privileged access to all sorts of information about women, this should actually produce counterevidence to any prejudice about *any* woman. If, because of my specific woman-friend, I have learnt that *not all women* are x-ing or y-ing, and if I then still reject a woman's testimony because I (on some level) believe that all women are x-ing and y-ing, this is epistemically illegitimate indeed. On the other hand, as was said earlier, prejudices often operate unconsciously – and sometimes despite the fact that I rationally agree with all sorts of counterevidence.

None of this yet explains why it is especially bad if I reject my friend's testimony – if at all, it suggests that it is especially bad for me to reject *any* woman's testimony on grounds of an unfair identity prejudice, precisely because I have a friend who is a woman. Also, whether being friends with someone constitutes access to counterevidence of a prejudice towards their social identity, is, again, an empirical question. In fact, assume briefly, that Angela is a pathological liar, and that generally, it would be far from epistemically illegitimate to ascribe a low degree of credibility to her testimony. Nevertheless, it would still be wrong for Bert to reject her testimony *on grounds of an identity prejudice against women*. Even if all existing evidence about Angela were in support of Bert's prejudice against women – assume that Angela does in fact tend to overreact and misinterpret other people's behavior towards her – this still would not render the prejudice itself or an invocation thereof epistemically legitimate. Put simply, the problem with being prejudiced is the prejudice, an a priori judgment made about something that can only be validated with respect to the empirical world.

Yet, we might wonder, is it only and simply the gathering of information and a rational weighing of reasons against the facts in support of those reasons that leads to the suspense of prejudice in real world cases? If I make friends with any member of a group that I am holding a prejudice against, isn't there something else at play that should lead me to reflect carefully on whether I am in any way prejudiced against that particular group? Perhaps we could point to the role that empathy plays in this regard, and arguably, empathy can play a decisive role in the epistemic process of receiving and evaluating someone's testimony. There is a lively debate evolving around the question of what exactly empathy is. However, for the present purpose, it should suffice to have a rough idea of what empathy *can do* rather than a precise one of what *it is*. According to Coplan (2011), empathy necessarily involves "perspective taking" and "perspective taking is an imaginative process through which one constructs another person's subjective experience by simulating the experience of being in the other's situation" (Coplan, 2011: 9–10). Let us further assume that this kind of perspective-taking can serve as a tool not to gather information, but rather counterbalance the way that information is presented to us through hegemonic narratives. If Bert were to take Angela's perspective from time to time, using all his imaginative power to try and simulate a woman's experience in a world that tells a very specific set of stories about what women are and what they experience, what they are supposed to be and what is expected of them, what they want etc., isn't it plausible to assume that this could push Bert to doubt prejudicial narratives? I do not want to make any assumptions about the epistemic quality or legitimacy of these doubts here, I just want to assume that these doubts *can* come to exist *qua* empathy. Now, having a friend who happens to be a member of some specific group may give me a better chance to empathize with their experiences and overcome my prejudice (by means of the frequency of occasion or by means of the depth of our conversations). However, it is far from clear if this creates an *obligation* to empathize in order to overcome my prejudices. More importantly for my purposes, even if there were such an obligation, this does in any way show how the unfair rejection of a friend's testimony based on the failure to discard my own prejudices is worse than my rejection of the testimony of any other (non-friend) member of the same specific group.

However, empathy may not only be a source for overcoming prejudicial beliefs, but may (to some extent) be the right kind of attitude to generally have when hearing another person testifying. Katharina Sodoma (2024) argues that particularly in cases of testimony of oppression, "*empathizing* with the speaker is a particularly promising way" of confronting that testimony (Sodoma, 2024: 186).[7] Testimony of oppression here means the testimony of an experience of oppression on behalf of the speaker, as in Angela testifying that she has been sexually harassed. But, as Sodoma herself convincingly argues, even though empathy has the advantage of enabling an understanding of the speaker's testimony or at least of the speaker's vulnerability, our understanding of a speaker's testimony should not be a condition for accepting the claim (cf. Sodoma, 2024: 199). Besides, it is not clear whether we have a special obligation to empathize when receiving testimony by our loved ones (more, or in a different way, than with the testimony of other people). Finally, even if such an obligation existed with regard to the testimony of oppression, this does not solve the problem at hand. Remember, that the intuition we aim to shed light on is that testimonial injustice in the context of intimate relationships, even if *not* a form of testimony of oppression, is still particularly bad.

Imagine another scenario, in which Angela tells Ernie and Bert over lunch that their boss made a mistake and that the numbers for the next quarter are all wrong. Ernie and Bert, due to being prejudiced against women (e.g., they believe that women are no good at math and have no clue how management works), do not believe her but instead suggest that she herself got the numbers wrong. Further, assume that also in this scenario, Ernie and Angela are only colleagues, Bert and Angela have been friends for six years. Even if this example is situated differently with respect to the vulnerability that Angela shares, it still seems that there is a significant difference in Ernie's and Bert's respective failure to believe Angela. Can we reasonably assume that a lack of empathy makes that difference? I am not sure. Nevertheless, I do believe that empathy points us in the right direction in so far as it is pointing to an affective dimension in cases of testimonial injustice at stake here.

7 Note that Sodoma's (2024) concern is somewhat more specific and targets the question how *privileged* hearers should confront testimony of oppression.

Let us recap what we have so far: Intimate relationships create (some kind of) special obligation on part of all parties involved; they warrant and arguably even command the *special* treatment of *specific* persons. In contrast, what Fricker is concerned with is an *equal* treatment of *all* persons in terms of awarding them the epistemic respect they deserve. Hence, there are two distinct questions to be addressed here: What do we owe each other as epistemic agents? And what do we owe our friends, lovers, or family members when engaging in epistemic practice? In the following, I attempt to appeal to recognition rather than to empathy for an answer. Let me explain.

1.2 Recognition, Love and Testimonial Injustice

In Paul Giladi's (2020) words, practices of testimonial injustice are "particularly harrowing", in so far as that they "rob a group or individual of their status as rational inquirers, thereby creating an asymmetrical cognitive environment in which they are not deemed one's conversational peer" (Giladi, 2020: 671). Giladi is explicitly exploring the concept of testimonial injustice from a recognition-theoretical perspective, an effort that recently is being undertaken by a growing number of authors (for example, see Congdon 2017; Hänel 2020; Giladi & McMillan 2023). The idea at stake is to not only conceive of testimonial injustice as objectifying or wrongfully discrediting a speaker, but further as the wrongful absence of some kind of (mutual) recognition obtaining between hearer and speaker. Let me briefly expand on this. Philosophical theories of recognition are manifold and diverse, going back at least to Rousseau, and occupying a special place in German Idealism within the works of Fichte and Hegel. In more recent times, Axel Honneth's (1995) *The Struggle for Recognition* takes Hegel's three forms of intersubjectivity (love, the law, solidarity) as a starting point to argue that we can only form an identity as a positive relation-to-self through the development of three corresponding modes of (intersubjective) self-realization: self-confidence, self-respect and self-esteem (cf. Honneth, 1995; Hänel 2020; Crawford 2021). Roughly, those three modes of self-realization via mutual recognition take place in different spheres having different implications: I acquire *self-confidence* in terms of some baseline stability of my own self in all my human neediness for love and affection within the realm of love and family, my *self-respect* is constituted on grounds of my being the part of a legal community that grants equal rights and respect to all of its members, my *self-esteem* comes from being part of a community in which my achievements, talents, accomplishments etc. are valued by fellow members. If

I can relate to myself in any of the ways outlined above, this will and must always happen intersubjectively; by means of being recognized by others that I in turn recognize. (cf. Crawford 2021). This is a very sketchy account of what is at the heart of several varying and complex debates, but I am positive that this sketch can suffice for my present purposes. Returning to testimonial (or, more broadly speaking, epistemic) injustice, recognition theory helps us to understand how we – as epistemic subjects, or as rational inquirers – are always (and only) an epistemic subject or a rational inquirer within a specific social context. I can only regard myself as an epistemic subject when I am part of some larger epistemic community; if my epistemic efforts in acquiring, refining, and communicating beliefs or knowledge of any kind within a community and if that epistemic community recognizes that my efforts exist and as such are, at least to a degree, valid. Being misrecognized, on the other hand, implies that there are very practical consequences (e.g., of not being believed), but also and crucially that any systematic misrecognition in this regard may make it hard to impossible for me to recognize myself as a capable epistemic agent. In short, epistemic injustice "undermines a positive relation-to-self through the denial of one's standing as a knower" (Jackson, 2018: 3).

The clear-cut distinction in between three different modes of intersubjective recognition as respectively enabling self-respect, self-confidence, and self-esteem may be debatable. I am not certain if it is plausible to assume that our lives operate in three distinct modes of intersubjectivity. Furthermore, a person's interaction with other people may involve more than only one kind of self-recognition at the same time. As Hilkje Hänel observes "a person's identity cannot be neatly split into three different parts" (Hänel, 2020: 7). Instead, let us assume that those three modes of self-recognition are all constitutively integral (even if somehow overlapping), and let us still hold on to the distinction in order to focus on *self-confidence*. This, in Honneth's 1995 treatise, is understood as being enabled via intersubjective recognition in contexts of care, that is, the context of love-relationships – of friends, lovers, siblings, or a parent-child relation.[8] Honneth, again building on Hegel, understands loving relationships as "conceptually and genetically prior to every other form of reciprocal recognition", or, to put it in Hegelian terms, as "the structural core of all ethical life" (Honneth, 1995: 107). Let me briefly expand on this, as it is exactly

8 This list is, by no means, intended to be exhaustive. What exactly qualifies as a loving relationship (and what should qualify as such) is an open question I do not intend to reply to in this essay.

these kinds of relationships that are at the heart of the question of my inquiry. Following Honneth, any loving relationship "prepares the ground for a type of relation-to-self in which subjects mutually acquire basic confidence in themselves" (ibid.). Recognition, here, is the constitutive element of love; precisely because the other's emotions are out of my control, I need to recognize both the independence of the other and of myself as well as the dependence of myself on the other in terms of my vulnerability to have my emotional needs fulfilled. The experience and expression of my emotions and emotional needs, possible only through the intersubjective experience of love, "helps to bring about, constitutes the psychological precondition for the development of all further attitudes of self-respect." (ibid.) To sum up: the mutual recognition involved in loving relationships is a prerequisite for whether a person will be able to positively recognize themselves in any other capacity or social scenario. Note how the appeal to recognition-theory also nicely shifts the emphasis with respect to the previously addressed intuition that we somehow owe our loved ones to empathize, that is, to try and take their perspective. I suggest that rather, what we owe is *the recognition* of the fact that our loved ones a) have their own, particular, and possibly unique perspective and that b) the perspective we ourselves have, has an impact on the perspective of the other. I will get back to that line of argument later.

Let me make two quick observations about the relation of basic (self-)confidence to intimate relationships before I finally return to the question of what happens if testimonial injustice occurs within that very sphere of an intimate relationship. Without delving much deeper into Honneth's account for now, I think it is fair to already ascribe a high degree of intuitive plausibility to what has been said so far. However, and as pointed out by other authors before me, it is slightly concerning how, for Honneth, the sphere of loving relationships, as one of family and privacy, seems to be confined to a social context that is somehow pre-political (cf. Connolly, 2010). Honneth does, after all, explore a relationship between power and recognition. The assumption that there is no political, social or cultural power operating at the level of intimate relationships and familial ties has been extensively attacked, especially by feminist scholars for what I hold to be very good reasons (cf. de Beauvoir 1953, Fraser 1989; Benhabib 1992). In fact, if political and social power relations had *no* impact on interactions in the sphere of love, friendship and family, my whole inquiry would be pointless, for then, it would make no sense to assume that testimonial injustice as a categorical, systematic practice of epistemic wrongdoing, which is rooted in the pertinence of identity prejudices stemming from social pathologies of marginalization, *even occurs* in the sphere of friendship, love or family. However, it so clearly does. For lack of space, I can only refer the doubtful reader to the fact that I am also drawing from personal experience. Further, I want to emphasize that just because recognition involved in intimate relationships is prior to all other kinds of recognition, we should not assume that this happens at an early stage in life and is somehow completed before a person goes on the quest for other kinds of self-recognition. It certainly is very likely that, if you do not experience love at all during early stages of your life, you thus fail to relate to yourself and the other in a way which enables a basic degree of (self-)confidence. Nevertheless, I doubt that this process can ever be quite complete over the course of a human life.[9] Certainly, even our most basic sense of confidence in ourselves, others, and the world can be interfered with at any point in our lives.

With this in mind, let's briefly return to Angela, Ernie, and Bert. To understand recognition as a vital human need and epistemic (self-)recognition as a prerequisite for a fully functioning epistemic agent, puts us in a better position to understand what happens to Angela when her testimony is being rejected. In general, testimonial injustice may generally interfere with an agent's self-respect in so far as that agent is unjustly excluded from an epistemic community that she has a basic right to be a part of. Testimonial injustice may also generally interfere with a speaker's self-esteem, that is, with the kind of (self-)recognition of an agent's achievements, talents, accomplishments or the like (cf. Giladi, 2018). But what happens between Angela and Bert happens within the realm of love, care and intimacy, and it violates Angela within the boundaries of a sphere that is supposed to foster her basic sense of (self-)recognition and thus the kind of self-confidence enabling Angela's *basic* ability to relate to herself and others in a positive way in *any* other social context.

Let us take a few steps back and reflect in more detail on Angela and Bert. It is plausible to assume that Angela *puts more weight on* what Bert, her friend, does or does not believe in general. Perhaps, when Bert does not believe *her*, this leads her to doubt her very experience of sexual harassment. She trusts Bert, she values Bert for being an intelligent and compassionate companion, and when Bert replies that she probably mistakes a harmless flirt for something it is not, this may lead Angela to question her very experience, her personality, and thus her epistemic sufficiency as an epistemic agent. In a nutshell, it is

9 To be sure, I am emphasizing that point with no intention to suggest that Honneth or other recognition theorists fail to recognize this.

very likely that Angela suffers from a greater loss of epistemic self-confidence when Bert does not believe her as compared to when Ernie does not. And, of course, ultimately this comes down to an empirical or psychological question. The extent of the damage a victim of testimonial injustice suffers from is most likely to differ from case to case, depending on a number of characteristics of the token case at hand. Yet, one of these characteristics is the relation between hearer and speaker. It presumably also matters whether and to which extent the speaker (systematically) has been a victim of unwarranted disbelief in the past. It matters how aware a victim of testimonial injustice is of the fact that her social surrounding feeds into identity prejudices towards members of her social group. Crucially, it presumably also matters very much how epistemically confident the speaker is in the first place.

This then, leads us back to Honneth and the idea that loving relationships are precisely what forge and foster self-confidence at some baseline level. Here, we are finally able to conceive of the distinct kind of badness involved in Bert's case not in consequential terms, but by reference to the characteristic structure of cases of testimonial injustice within the sphere of intimacy. Critically, the point with Bert's case is neither that someone like Angela *is very likely to suffer* from a decrease in self-confidence, nor that Angela is subjected to *a higher risk of loss in self-confidence*, even though both points are very valid. Rather, the point is that what happens in Bert's case does not only interfere with Angela's self-confidence but that it happens in not just any social (and/or epistemic) context, but in the very place that is supposed to somehow form *the core* of social (or, for that matter, epistemic) interaction. Hence, this case is not only about self-confidence, but about a basic sense of the confidence in others, the world, and ourselves. If, as I briefly pointed out above, it is true that the process of acquiring a basic sense of confidence via recognition in the sphere of love has to be reinstated and reaffirmed over the course of our lives, it becomes clear how harrowing the occurrence of testimonial injustice is in the context of intimacy. Then, Bert's rejection of Angela's testimony might come down to something like a rupture, a break of Angela's basic confidence in herself and her social surroundings; that is, a genuine interference with her basic sense of (epistemic) agency.

2. A Special Kind of Badness

As we have seen in the previous section, it is plausible to assume that all cases of testimonial injustice may be understood as involving the failure to award a speaker with the right kind of recognition. In the case of testimonial injustice occurring in the context of intimate relationships, this comes down to an interference with a person's (self-)confidence in the most basic and yet critical kind of way. In the following, I want to delve deeper into that idea and motivate the idea that we should understand cases of testimonial injustice in the sphere of intimacy as cases of testimonial betrayal. This, I hope, serves to clarify the harm done to not only the speaker whose testimony is being rejected, but also to the relationship between hearer and speaker. More specifically, I argue that the kind of thick trust in place in intimate relationships serves as somewhat of *a guarantee* that a person will be recognized as an individual in all their particularity. A failure to award that kind of recognition to a friend or lover, mother, father, child, or comrade, especially if tied to an identity prejudice about the already marginalized group that person is a member in, is not just an epistemic failure and likely to destroy their self-confidence, but constitutes a genuine case of betrayal of the relationship in place. Let me provide some more detailed justifications for these claims in the following.

2.1 Introducing Testimonial Betrayal

Wanderer (2017) argues that the decisive difference between testimonial injustice and its instantiation in the context of intimate relationships is the thick trust relationship that obtains between parties in case of the latter. In thick trust relationships, parties seek each other's approval or recognition, and a failure to accord this recognition, after Wanderer, constitutes a case of *testimonial betrayal*.[10] Let me explain. If we apply Wanderer's formula to our exemplary case at hand, this means that the significant difference between Bert's and Ernie's respective rejection to Angela's testimony manifests itself in differing degrees (or, as I argue later, kinds) of trust at play. That is, the thin stance

---

10 To be sure, Wanderer remains cautious to draw any definitive conclusions in his treatise. He draws a rough sketch of a concept, and then provides some tentative arguments about why this concept should be treated as a variety of testimonial injustice. I take these insights as a starting point to develop Wanderer's ideas further. My enquiry is not intended as a critique of Wanderer.

of trust adopted towards each other in Angela's and Ernie's case critically differs from the thick stance of trust adopted between intimates, such as Angela and Bert (cf. Wanderer, 2017, p. 36). Put differently, what Angela and Ernie owe to each other is determined by their capacities as hearer or respectively knower, whereas Angela and Bert's are additionally bound by the particular friendship obtaining between them. Wanderer argues that in an intimate relationship like this, "parties seek each other's approval, often including recognition of their opinions on matters arising in the course of their interactions" (Wanderer, 2017: 37). In cases of testimonial injustice, a failure to accord the recognition sought is turned into a rejection of the requested approbation: "It is not just that one party does not provide the recognition that the other party seeks, but that the one party throws the request for recognition back to the other party" (Wanderer, 2017: 37). Angela has made a humiliating experience and attempts to communicate that experience to her friend Bert. She thereby seeks his recognition, not so much of an opinion of hers, but of the fact that she experienced an act of suffering, of humiliation; in short, a moral wrong. Bert is in the exclusive position to assure Angela that her feeling of suffering and humiliation bears justification, that her perspective on the world is not thwarted, that her reaction corresponds to the (moral) character of what happened to her. The problem now is that it is far from obvious if this applies exclusively to intimate relationships characterized by thick trust. The kind of disrespect featuring in *all* cases of testimonial injustice can be plausibly understood as a rejection of the request to be recognized in the right way, that is, as the full moral person one is, as a rational enquirer and testifier of knowledge. Yet, following Wanderer, when Bert rejects Angela's testimony, she is not only wronged in her status as a person, but "his failure to believe her is experienced by her as a rejection of their relationship" (Wanderer, 2017: 36). Crucially, what is thus being betrayed is the intimate relationship between hearer and speaker itself.

We are now finally in the position to give the distinct badness of Bert's case a name: testimonial betrayal. But the name itself does not yet grant a sufficient explanation of how precisely the badness evolves. Simply to say that cases of testimonial betrayal are worse than other cases of testimonial injustice because they are cases of betrayal only begs the question. Wanderer repeatedly speaks of the "sting of betrayal" that is felt or experienced (cf. Wanderer, 2017: 36–37). This suggests that what makes betrayal bad, is that she who is betrayed feels bad, or suffers extensively, or is being injured. And this in turn suggests that what is so particularly bad about testimonial betrayal is nothing beyond the fact that the victim suffers from an even greater injury – a double sting –

than in other cases of testimonial injustice. This might well be true, but likewise dissatisfying as an answer to our problem.

Further, the intuition that Bert's case of non-belief is worse than Ernie's case may still hold without making any assumptions about Angela's respective feelings. In the previous chapter, I have introduced the idea that when Bert fails to believe Angela, he is not only failing to award her the right kind of recognition, but by means of this, interferes with what is at the core of her (self-)confidence. Also, I have already suggested that we should conceive of the badness of this interference not in consequential terms and that what matters here is not the degree to which Angela's (self-)confidence is *actually* diminished. Instead, I believe that the badness at stake lies in *the interference itself*; that is, in the simple fact that this interference takes place. To strengthen this claim, I believe it is helpful to have a more detailed look at the concept of betrayal. After all, betrayal is a strong word that carries moral weight. And, although Wanderer equips us with the suitable kind of terminology, his paper leaves open the question of what is particularly bad about betrayal; a task I turn to next.

2.2 The Badness of Betrayal[11]

Avishai Margalit (2017) provides a book-length treatment of the phenomenon of betrayal. My inquiry cannot afford an exhaustive discussion of Margalit's account. However, I think that even a rough sketch of Margalit's conception of betrayal suffices to get a better grasp of why testimonial betrayal is a form of betrayal as well as of the badness attached to it. Margalit starts by investigating the effect of betrayal to ultimately infer the characteristics of betrayal. The effect of betrayal centers around the assumption that "it is the injury to the relationship that makes it betrayal" (Margalit, 2017: 83). This injury is later spelled out in more precise terms as the undermining of thick human relationships (cf. Margalit, 2017: 84–94). What is meant here with the notion 'thick relationship' is roughly what I refer to as intimate relationships: relationships of friendship, love and so on. However, Wanderer's notion of 'thickness' is a rich

11 In this paper, I argue for the intrinsic badness of (testimonial) betrayal and assume that all cases of betrayal are intrinsically bad because they undermine the meaning of the relationship between betraying and betrayed party. Naturally, this does not rule out that there may be cases of betrayal that are somehow not-so-bad, or even good in terms of their consequences. If this sounds counter-intuitive, think of whistleblowing. Betrayal being bad hinges on the premise that the undermining of a (good) relationship is bad, which I take for granted here.

one, for it tracks the distinct character of intimate relationships. I shall illuminate the mystifying notion of thickness in a moment, let me first show what I take to be the core of the argument. Betrayal, following Margalit, has the effect of undermining a thick relationship insofar as it terminates or erodes a relationship (cf. Margalit, 2017: 88). Yet, what is being eroded or terminated is not necessarily the relationship itself, but first and foremost its thickness. Consider a case of betrayal amongst family members: If a son betrays his father, the general relationship between the two of them will not cease to exist, they are still father and son. However, the relationship may practically cease to exist in terms of a thick (and not merely a trivial) relationship. They might still be father and son, but no longer be involved with each other in any of the relevant ways in which fathers and sons can be involved with each other (caring for each other, trusting each other, being a part of each other's life and so on). What about our case? After Bert fails to believe Angela that their boss sexually harasses her, Angela and Bert may still see each other at work, they may continue to go to lunch together and since they are part of the same circle of friends, even continue to spend their late nights in the same pub. From an outsider's perspective, the relation between the two still displays all the typical features of friendship. Nonetheless, from Angela's perspective, it might not make much sense to call Bert her friend any longer. So much for the *effects* of betrayal: it undermines the thickness of a relationship. This leads Margalit to argue that the characteristic feature of betrayal itself, is the undermining of *the meaning* of a thick relationship; i.e., the specific forms that thick relations take. As he puts it, thick relations "are relations under a description: my sister, my friend […] The meaning of each of these relations has a different role in my life and in the lives of those with whom I stand in such a relation. Undermining the meaning of thick relation is undermining the specific description the relation goes under" (Margalit, 2017: 88). Crucially, such descriptions do not designate the specific conventional type such as, say, friendship, but the specific token of a thick relation, such as the very particular friendship between Bert and Angela. As Margalit correctly observes, "in thick face to face relations we have specific people in mind and we don't have to idealize them, for we care about them in the specific way they are" (Margalit, 2017: 79). Furthermore, Margalit identifies a normative sense of betrayal: not only is it the case that thick relations may be terminated, but this happens justifiably so, for betrayal provides the betrayed with a reason to reevaluate the meaning of the thick relation between her and the betrayer (cf. Margalit, 2017: 92). This is also what Wanderer seems to have in mind, when he states that "failures to live up to the expectations of that thicker relationship may well leave one party questioning whether such a thicker relationship was actually in place in the first place" (Wanderer, 2017: 37) Recall further that in this regard, Wanderer speaks of relations forged by thick *trust*. Margalit, however, does not speak of trust, but of a "reevaluation of the worth and significance of the relation" (Margalit, 2017: 94). This brings me to my next question: What kind of worth and/or significance is (ideally) shared by all intimate relationships? A high degree of trust may be part of this worth and significance, but, as Margalit argues, thick trust is simply the manifestation of a thick relation (cf. Margalit, 2017: 84). That is, the fact that there is thick trust simply stems from there being a thick relation, and there being a thick relation manifests itself through thick trust. This makes sense. Taking a look at real-world examples of, say, friendship, it is in fact almost impossible to determine what came first: the friendship or the kind of thick trust that obtains between friends?

Yet, following Margalit, there is another significant feature shared by all thick relationships. A feature that serves as a better demarcation criterion than trust and also creates the background against which thick trust and thick relations can emerge and grow in the first place. The thickness of thick relations is, after Margalit, best accounted for by a sense of *belonging*. More specifically, he distinguishes between relations oriented by belonging and relations oriented by achievement. The latter is what applies to most relations obtaining in a person's social life. The former is what applies exclusively to thick relations such as intimate relationships. Corresponding to these two kinds of orientation, persons evaluate each other in a context of an interaction either in terms of achievement or in terms of belonging (cf. Margalit, 2017: 84). Whereas in a relation oriented by achievement, people's engagement with each is governed by what they are trying to achieve. In comparison, the interaction within a relation oriented by belonging is characterized by a special kind of bond between specific individuals. My inquiry does not permit for a complete defense of Margalit's conception – but note how this nicely maps with some of the insights we drew from recognition theory in the previous chapter. Recall that here as well, we encountered the idea that recognition is something we both owe and need from each other, but that is realized in different forms, according to the social sphere in which we encounter each other. Margalit's thick relationships then correspond to the sphere of love and intimacy that enable our most basic sense of confidence *qua* recognition. Margalit's achievement-oriented relationships correspond to the kind of recognition that enables us to develop self-esteem. However, recognition theories are not about a simple exchange of attitudes owed, but about a necessarily intersubjective normative outlook. That is, my self-recognition is only *enabled through* the recognition of the other and vice versa. This adds an important layer to Margalit's way to distinguish between achievement- and belonging-oriented relationships.

If you are the teacher and I am the student, we evaluate – and recognize – each other in terms of my learning something and you teaching me something. What we owe to each other is determined by our respective capacities within the framework of the interaction. That is, I am obligated to listen to you in your capacity as my teacher. Of course, this also hinges on the extent to which you live up to what a teacher should or should not do by means of being a teacher. Following Margalit, I evaluate you in terms of achievement. Intimate relationships function differently. At first sight, we may indeed evaluate each other in our respective capacities as each other's friend. That is, the general idea of friendship might implicate some general obligations for all parties involved. However, these obligations are constituted less by the general idea of friendship but by *our very own* friendship. It is somewhat less clear what to expect from a friendship than what to expect from, say, a teacher-student-relationship.[12] In other words, there is no clear idea of what is to be *achieved* in the framework of a friendship, except the maintaining of the specific sense in which you and I belong to each other as the very persons we are.[13] If I betray you, I undermine the specific way in which you and I stand towards each other as the specific persons we are; that is, I undermine the specific way in which we *belong to each other*. The point is: All friendships are different, thus, we should recognize the very specific and unique way, in which you and I relate to each other, being in turn enabled by the particular individuals that you and I respectively are.

Now recall that Wanderer claims that what it means to be in intimate relationships and to engage in testimonial exchange is to seek each other's approval or recognition. How then does the form of recognition sought here differ

12 For a more thorough exploration of friendships and romantic relationships see Hänel & Jenkins (2024), who convincingly argue that there is no proper grounds for holding up a clear-cut distinction between the two; and that the shared projects of intimates are defined by their particular and individual relationship rather than by any preconceived 'form' of the so-called 'type' of relationship.

13 This does not rule out the possibility of shared goals within the framework of an intimate relationship. The point rather is that if two intimates have shared goals or try to achieve something together, this would be a *distinctive* characteristic of this very intimate relationship.

from the way in which persons generally seek each other's recognition within the context of giving testimony. Why does the former open up the possibility of betrayal, whereas the latter does not? This is my tentative reply to this question: Testimonial injustice per se critically involves a kind of misrecognition on part of the hearer. In the case of a failure to award another epistemic agent the recognition they deserves, this is an unjust evaluation of them in terms of achievement as an epistemic agent. Yet, in the case of intimate relationships, every single instance of testimonial exchange is embedded in the thick relation between two persons who belong to each other in a specific way. Recognition is therefore not only sought for the capacity to gain and convey knowledge, but for a) the very individual person that the testifier is[14] and b) the specific relationship between hearer and speaker. What Bert owes to Angela is not merely the recognition of her testimony in terms of her being a person with full moral status, including her status as a knower, but in terms of a) the very specific person Angela is and is known to Bert and b) the bond of the relationship that both of them mutually share. Cases of testimonial exchange within the context of intimate relationships warrant not only the recognition of the speaker as an epistemic agent, but of the specific individual to whom the hearer stands in a thick relation to and of that relation. The kind of thick trust in place in intimate relationships, that is, the way, in which we belong to one another[15], is a way to trust in the fact that we will be recognized as the individual people we are, and in all our particularity. A failure to award that kind of recognition is not just an epistemic failure, or likely to destroy their self-confidence, but constitutes a case of betrayal.

2.3 The Badness of Testimonial Betrayal: Bert's Case Revisited

The upshot of my brief excursion into the characteristics of betrayal is the following. Margalit (2017) argues that betrayal undermines the meaning of thick human relationships. What makes betrayal bad can thus not only be accounted

14 I do not mean to imply that people have fixed identities, I simply adopt some common-sense, but sloppy view on how we tend to assume that another person is a specific and distinguishable person. My point is more about every person being particular and special, this still holds true if we assume that identities change over time or all the time. And even if there is no such thing as an identity, I think that some idea of an identity is still presupposed in everyday-life kind of interactions.

15 Perhaps, a better way to put this would be not so much to say that 'belong to another' but instead that we 'belong to the relationship' in place.

for by appeal to the injury of the betrayed party, but by the fact that she is given a good reason to reevaluate the significance and worth of the relationship. How can this then help to understand testimonial injustice within the context of intimate relationships. Following the discussion above, I hold that to be prejudiced against someone and to reject their testimony on the grounds thereof is bad in a different way when it happens within the context of human relations oriented by belonging than it is within the context of relations oriented by achievement. In general, the fact that human beings engage in the social practice of testimony creates relations, which are oriented by the achievement to development a huge body of knowledge that is shared by all parties involved. When Angela talks to Ernie, they evaluate each other in their capacities as knowers who can potentially achieve knowledge acquisition and sharing. When Ernie negatively evaluates Angela in her capacity as a knower based on an unfair identity prejudice, this is a manifestation of an unjust evaluation of Angela with regard to said achievement; thus, the unjust exclusion of Angela from the pooling of knowledge. In comparison, when Bert fails to believe Angela, they stand in two kinds of relation to one each other. The first relation is their relation as being both epistemic agents involved in the social practice of testimony. Thus, the injustice manifest in Ernie's rejection of Angela's testimony likewise obtains in the case of Bert and Angela. However, the second relation is their thick relation of friendship. Here, the problem constitutes a case of betrayal as specified above.[16] Angela tells Ernie and Bert the same story in the exact same words at the same time, both do not believe her, both reject her testimony based on the same prejudice. But when Angela is speaking to her friend Bert, she is speaking to him as her friend. The kind of recognition owed, or to be reasonably expected in the realm of intimacy is not tied to any specific function or capacity, but rather to the recognition of the independence and dependence of the individuality, the feelings, attitudes, and characteristic features of the other and the bond shared. [17] Bert, even if unconsciously so,

16 One way to conceive this would be that the testimonial injustice at hand manifest itself via the unwarranted way, in which Bert does not believe Angela, the betrayal at hand rather can be traced back to a *belief* Bert has *about* Angela when he is prejudiced towards her. For a convincing defence of the view that there are obligations in terms of the beliefs persons have about one another see Basu (2019).

17 This is not supposed to imply that we do not owe respect to all people in terms of a recognition of their (infinite) specificity and particularity. I believe we do, but I cannot argue for this within the scope of this paper. For a convincing argument in this regard, see Giladi 2020.

betrays that very bond in between Angela and him. Margalit points out that one canonical form of conceiving betrayal comes in the ternary relation of "A betrays B to C", roughly corresponding to something like "handing" someone "over to the enemy" (Margalit, 2017: 70–71). What happens in between Bert and Angela certainly does not come down to any literal version of that. But, roughly speaking, Bert betrays the friendship *to* the prejudice he holds against women. Since the prejudice is only in place because of a structural form of marginalization, I think we can agree that Bert, in a way, hands his marginalized friend over to the enemy as the marginalizing party.

The overall aim of this inquiry was to account for the perceived moral difference of Bert's and Ernie's respective treatment of Angela's testimony. By now, it should be clear that testimonial injustice in the context of intimate relationships is neither more unjust nor does it necessarily feature a higher extent of epistemic illegitimacy. Instead, cases like the one discussed bear an additional distinct kind of badness, that is, the badness of the betrayal. At this point, one might object that the badness of Bert's case cannot be traced back to the badness of one phenomenon, but rather to the coincidence of two phenomena, ***viz.*** testimonial injustice and testimonial betrayal. However, this should not be understood as a coincidence. Instead, I believe that the fact that testimonial injustice *even* occurs within the sphere of love, friendship and other forms of intimacy just shows how asymmetrical power structures and the systematic exclusion of marginalized voices operate at every level of social interaction. What testimonial injustice poses a threat to is not only our being within an epistemic community, but to our relations with loved ones. This being said, the precise relation in which testimonial injustice and testimonial betrayal stand to each other, certainly needs more conceptual clarification and leaves room for further work.

3. Conclusion

As Wanderer (2017) puts it, "the actual social practice of testimony does involve socially-situated agents, and this means (...) that they stand in varying relations of intimacy and acquaintance with each other" (Wanderer, 2017, p. 37). The preceding discussion hopefully enriches the understanding of how varying relations of intimacy feature in cases of testimonial injustice. More specifically, I shed light on the instantiation of testimonial injustice within the sphere of thick intimate relationships. In this regard, I motivate and account for the intuition that cases like these critically exhibit more badness than other cases of testimonial injustice. The distinct intrinsic badness featuring in all relevant cases, or so I argue, is better captured by understanding them as cases of what Wanderer (2017) calls testimonial betrayal. I find that it is in fact the badness of betrayal that renders cases of testimonial injustice in the context of intimate relationships particularly bad. If testimonial injustice occurs within the framework of an intimate relationship, this undermines the significance and worth of the intimate relationship itself, interfering with the sphere of life that fosters and maintains our basic sense of confidence in ourselves and the other. The kind of recognition owed in intimate relationships is the recognition of the relationship itself and all parties involved, with the particular and unique features of that relationship. A failure to believe an intimate based on an unfair prejudice about their identity amounts to a failure to recognize your intimate as the person they are as well as a failure to recognize the relationship in place. The crucial difference between Frickerian cases of testimonial injustice and cases of epistemic injustice within the context of intimate relationships (and the explanation for the intuition that the latter are in some way worse) is thus first and foremost this: the latter undermines the meaning of a relationship, which simply does not obtain within the framework of the former.

References

Baier, Annette (1986): Trust and Antitrust. Ethics, Vol. 96 (No. 2): 231-260.

Bailey, Olivia (2018): Empathy and Testimonial Trust, Royal Institute of Philosophy Supplement 84: 139–160.

Barker, Simon, Crerar, Charlie, & Goetze, Trystan S. (2018): Harms and Wrongs in Epistemic Practice. Royal Institute of Philosophy Supplement, Vol. 84 (No.1): 1–21.

Basu, Rima (2019): What We Epistemically Owe To each Other, Philosophical Studies, Vol. 176 (No. 4): 915-931.

de Beauvoir, Simone, (2009 [1949]): The Second Sex, Vintage Books, London.

Benhabib, Seyla (1992): Situating the Self: Gender, Community, and Postmodernism in Contemporary Ethics. New York: Polity.

Christensen, David (2007): Epistemic Self-respect. Proceedings of the Aristotelian Society, Vol. 107 (No. 3): 319–337.

Coplan, Amy (2011): Understanding Empathy: Its Features and Effects'. In Amy Coplan and Peter Goldie (eds.), Empathy: Philosophical and Psychological Perspectives (Oxford: Oxford University Press): 1–18.

Crawford, Lindsay (2021): Testimonial Injustice and Mutual Recognition, Ergo: An Open Access Journal of Philosophy 7. doi: https://doi.org/10.3998/ergo.1128

Davis, Emmalon (2016): Typecasts, Tokens, and Spokespersons: A Case for Credibility Excess as Testimonial Injustice, Hypatia Vol. 31 (No 3): 485–501.

Dotson, Kristie (2011): Tracking Epistemic Violence, Tracking Practices of Silencing, Hypatia Vol. 26 (No 2): 236–257.

Fraser, Nancy (1989): Unruly Practices : Power, Discourse, and Gender in Contemporary Social Theory, University of Minnesota Press.

Fricker, Miranda (2007): Epistemic Injustice: Power and the Ethics of Knowing. Oxford University Press.

Fricker, Miranda (2012): Group Testimony? The Making of a Collective Good Informant, Philosophy and Phenomenological Research, Vol. 84 (No. 2): 249–276.

Giladi, P. (2018): Epistemic Injustice: a Role for Recognition? Philosophy and Social Criticism, Vol. 44 (No. 2): 141–158.

Giladi, P. (2020): The Agent in Pain: Alienation and Discursive Abuse, International Journal of Philosophical Studies Vol. 28 (No 5): 692–712.

Giladi, P. & McMillan, N. (eds.), (2023): Epistemic injustice and the philosophy of recognition New York, NY: Routledge Taylor & Francis Group.

Govier, Trudy, (1993): Self-Trust, Autonomy, and Self-Esteem, Hypatia, Vol. 8 (No. 1): 99 – 120.

Grasswick, Heidi (2018): Understanding Epistemic Trust Injustices and Their Harms. Royal Institute of Philosophy Supplement, Vol. 84: 69–91.

Harris, Paul L. & Koenig, Melissa H. (2007): The Basis of Epistemic Trust: Reliable Testimony or Reliable Sources? Episteme, Vol. 4 (No 3): 264–284.

Hawley, Katherine J. (2014),:Trust, Distrust and Commitment, Noûs, Vol. 48 (No. 1): 1-20.

Hawley, Katherine J. (2014): Partiality and prejudice in trusting, Synthese, Vol. 191 (No. 9): 2029–2045.

Hawley, Katherine J. (2017): Trust, Distrust and Epistemic Injustice. In: I. J. Kidd, J. Medina & Gaile Pohlhaus (eds.), Routledge Handbook of Epistemic Injustice. Routledge.

Hayakawa, Seisuke & Miyahara, Katsunori (2024): Empathy through Listening, Journal of the American Philosophical Association (2):1-16.

Hazlett, Allan (2017): On the Special Insult of Refusing Testimony. Philosophical Explorations, Vol. 20 (No. 1): 37–51.
Hänel, Hilkje C. (2020): Hermeneutical Injustice, (Self-)Recognition, and Academia, Hypatia Vol. 35 (No 2):1-19.
Hänel, Hilkje C. (2024): Willful testimonial injustice as a form of epistemic injustice, European Journal of Philosophy (1):183-201.
Hänel, Hilkje C. & Jenkins, Katharine. (2024): On Relationships. Zeitschrift für Ethik und Moralphilosophie 7: 459–486. 10.1007/s42048-024-00192-3.
Honneth, Axel (1995 [1992]): The Struggle for Recognition – *The Moral Grammar of Social Conflicts.* Translated by Anderson, J., Polity Press.
Jackson, Debra L: (2018): "Me Too": Epistemic Injustice and the Struggle for Recognition, Feminist Philosophy Quarterly 4 (4). https://doi.org/10.5206/fpq/2018.4.6231.
Lowe, Dan (2024): Testimonial Injustice and the Ideology Which Produces It, American Philosophical Quarterly 61 (3): 215–231.
Malcolm, Finlay (2018): Testimonial Insult: A Moral Reason for Belief? Logos and Episteme, Vol. 1: 27–48.
Margalit, Avishai (2017): On Betrayal. Harvard University Press.
Marušić, Berislav (2025): Interpersonal Reasoning: A Philosophical Psychology of Testimonial Trust. European Journal of Philosophy, 33 (2): 531–549.
McCraw, Benjamin (2015): The Nature of Epistemic Trust. Social Epistemology, Vol. 29 (No. 4): 413-430.
Origgi, Gloria (2012): Epistemic Injustice and Epistemic Trust. Social Epistemology Vol. 26 (No. 2): 221–235.
Perrine, Timothy (2023): Prejudice, Harming Knowers, and Testimonial Injustice, Logos and Episteme 14 (1): 53–73.
Pettigrew, Richard (forthcoming): What is the characteristic wrong of testimonial injustice? Philosophical Quarterly.
Sodoma, Katharina Anna (2024): Testimony of Oppression and the Limits of Empathy, Ethical Theory and Moral Practice 27 (2): 185–202.
Wanderer, Jeremy (2017): Varieties of Testimonial Injustice. I. J. Kidd, J. Medina & Gaile Pohlhaus (eds.), Routledge Handbook of Epistemic Injustice. Routledge.
Warman, Jack (forthcoming): Insult and Injustice in Epistemic Partiality, Journal of Value Inquiry:1-2.
Wilholt, Torsten (2013): Epistemic Trust in Science. British Journal for the Philosophy of Science, Vol. 64 (No. 2): 233–253.

Zagzebski, Linda (2003): Epistemic Trust. Philosophy in the Contemporary World, Vol. 10 (No. 2): 113–117.

What (Can) Students Know?

Epistemic injustice, recognition, and German schools

Peimaneh Yaghoobifarah

Children typically spend at least ten years of their lives attending school. During this time, school acts as the central place for them to acquire knowledge, conceptualize their social experiences through appropriate terms and make their own perspectives audible outside the family. They thus find themselves in the middle of the dichotomy between teaching and learning; surrounded by multiple collective knowledge resources that form the basis of this dichotomy. Yet children in general, and pupils or students in particular, are largely overlooked in the discussion of epistemic injustice initiated by Miranda Fricker (2007). Fricker's theory primarily focuses on marginalized adults who, in the continuity of structural injustices, are also prevented on an epistemic level from being recognized as knowers and from making their experiences intelligible for themselves and others. However, if we take into account the fact that children collectively become the target group of injustices, for example through physical and psychological violence, child poverty or child pornography, the question arises as to whether they cannot also be understood as a social group in their own right, who are important as subjects for theories of epistemic injustice on the basis of their group membership.

Based on the assumption that children, or in the school context, students[1], can be understood as a separate social group, I propose in this paper that students, as a result of their studenthood, can experience both testimonial and

1 For the purposes of this study, students are defined as all learners between the ages of 10 and 18. Normally, this age range includes the transition to secondary schools up to the potential university entrance qualification. The consideration of epistemic injustice among younger learners of primary school age would also be relevant in the fields of education, the healthcare system, and in the reporting of violent situations, but cannot be sufficiently addressed in this paper.

hermeneutical injustices. This marks an important distinction of multiply realizable epistemic injustice; including one that is not considered by Fricker. On the one hand, students, who are marginalized because they are marked as different from the dominant culture due to, for example, racialization or religious affiliation, can be treated in epistemically unjust ways because of their respective marginalization. This is a form of epistemic injustice well researched but not yet applied to students. On the other hand, students are additionally affected by epistemic injustice qua being students. This is a form of epistemic injustice that goes beyond the framework Fricker provides. It is important to consider both categories on the basis of which epistemic injustice can occur because it highlights nuances in the situatedness of students as epistemic subjects. *All* students can experience epistemic injustice, but they are not all exposed to it to the same extent. Furthermore, aspects of Fricker's theory show that the epistemic harms that subjects suffer as a result of their marginalization or their student status can adequately be regarded as a form of misrecognition. Accordingly, in cases of epistemic injustice, students are not recognized as full subjects because their status as knowers is restricted in social interactions. The connection to theories of recognition is particularly important because it reveals the structural level of epistemic injustice, which is not sufficiently taken into account in Fricker's own discussion.

To substantiate the central claim of this paper, I will begin to illustrate how marginalized students in Germany can experience epistemic injustice and incidents of misrecognition using examples of race and religious affiliation. Yet, a closer look at the school environment suggests that misrecognition is inherent in the school system and that students can therefore experience incidents of misrecognition not only through their marginalization but also through their status as children (or students). At this point, the question arises as to whether there can be epistemic injustice that does not necessarily occur through marginalization, but rather on the basis of being a student. By including students, or children, as a separate social group, new possibilities open up for conceptualizing not only the epistemic status of children more clearly, but also their general social status, and thus embedding it in the discourse on social justice. Finally, I consider what specific changes can be made to the institution of school in order to counteract the epistemic injustices and misrecognition that are continuously reproduced there. To this end, principles for a more epistemically just school are proposed that rethink outdated school patterns, create new spaces for recognition and epistemic justice, and suggest a concept of epistemic modesty that does justice to the epistemic potentials of children and adults alike.

1. Dealing with (marginalized) students in German schools

"Education is less a solution than a problem in its own right,"[2] writes sociologist and educator Aladin El-Mafaalani in his book *Mythos Bildung* (2020: 55). El-Mafaalani is referring here to the widespread but empty promise of equal opportunities in the German school system. The myth of equal opportunities suggests that existing social inequalities do not have an impact in schools, or that schools can even absorb and compensate for them. Instead, performance and effort are the keys to educational success (cf. El-Mafaalani 2020: 55). The fact that this assumption is more illusion than reality is hardly disputed in critical educational research (cf. Weber 2005; Fereidooni 2011; Karabulut 2016; Helmchen 2019; El-Mafaalani 2020). "The education system does not offer equal opportunities – even for equal performance" (El-Mafaalani 2020: 57). Reasons for the asymmetrical relationships in German schools are outlined below using the example of the school as an ideological state apparatus, as an educational institution, and teachers as one of the main pillars and mediators of the construction and reproduction of norms and discrimination in everyday school life.

1.1 Ideological: The naturalization of a capitalist order through school

The naturalization of differences and inequalities (Rommelspacher 2011: 26) also takes place in German schools, for example, when religious symbols such as headscarves are banned for teachers on the grounds that state institutions must reflect the separation of religion and state, but allowing crosses to remain hanging in courtrooms and ministries (cf. Spiegel 2021). Or when some children's lunchboxes are devalued and stigmatized in the schoolyard, while other eating habits are considered normal (cf. Hirose 2011). Often, it is not only characteristics and attributions of individuals that are used to mark them as different. Such marking can also take place on the person themselves. For example, when people are singled out in the classroom and portrayed as different because of their appearance or religious affiliation. Othering as a problem of

2 Unless explicitly stated, all translations of German are by the author of this text.

recognition also affects social participation and material conditions. Othering is a process that is related to social and societal relationships of dominance and subjugation and distributes social positions and the associated access to social resources and privileges (cf. Riegel 2016: 58).

It is important to emphasize here that these exclusions are always related to the social value of people. This means that all injustices and their associated ideologies, such as fascism, racism, or sexism, which take place on the basis of social groups, must always be located in the context of capitalism.

Thus, the asymmetrical distribution of socioeconomic resources is legitimized by the naturalization of social inequalities and the implied social value of people. Louis Althusser describes how schools, analogous to other state institutions, contribute to maintaining the central mechanism of exploitation and oppression in capitalism: "In other words, school [...] teaches 'skills', but in forms that ensure *submission to the dominant ideology* or mastery of its 'practice'" (Althusser 1970: 112). By qualifying school learning through subject-specific content taught in class, the 'rules' of interaction with one another, and the consolidation of social status not only for the reproduction of their labor power, "but also simultaneously a reproduction of their submission to the rules of the established order," the ideological order of capitalism is directly reproduced (Althusser 1970: 112). The ideological order of the school here means:

> Which norms, values, and attitudes students [...] adopt, but also how the knowledge imparted is applied and how everyday interactions in school are structured, all of this is controlled by open and latent mechanisms in the school, since the school rewards conformity and, if necessary, enforces it with [...] psychological violence. (Brandmayr 2017: 188)

Of course, the way in which a capitalist society—and, consequently, schools as preparation for this society—is not based on solidarity, participation, and the fair distribution of goods, but rather on the exploitation of entire social groups and the accumulation of wealth is not an official guideline that schools use. Rather, following Brandmayr, ideological values are primarily conveyed in schools through practices: Through the individualization of learning success, the idea of equal opportunities, the appeal to efficiency, personal responsibility, and self-optimization, the teaching of rules of conduct, and the hierarchization of content, "formal and informal ways of presenting what is possible, desirable, or prohibited" are established (Brandmayr 2017: 199). Those who do not conform to these practices are very likely to fail at school. Through the ideology conveyed in school, students are thus taught implicitly – rarely explicitly – that their social value depends on the production of their labor power and that their respective labor power is already predetermined by their social status within an order. The reality of a social order based on exploitation is naturalized and normalized (cf. Brandmayr 2017: 189). Whether students exploit or are exploited after graduating from school is therefore largely determined by the framework conditions of the school.

Althusser (1970) also describes how the subject is integrated into "the economic relations of production" (Brandmayr 2017: 189). From the moment they enter kindergarten, children are assigned to social classes and integrated into the social order between the state apparatuses of school and family. This order becomes visible at the latest when they reach middle school age: "At around the age of 16: an enormous mass of children 'fall' into production: the workers or small farmers. Another part of the school youth continues on" (Althusser 1970: 128). Who 'falls' and who is allowed to remain in the state apparatus of the school is, of course, not determined by chance. Althusser continues: "Every group that 'falls' along the way is practically imbued with the ideology that corresponds to its role in class society: the role of the exploited" (Althusser 1970: 128).

The place that each person occupies in the social order is also conveyed in terms of its content. According to Christine Riegel: "Ideas of normality and knowledge dominate, which are subject to the ethnocentric, racialized, heteronormative, middle-class, and ableist orders of meaning and difference of Western capitalist relations" (Riegel 2016: 85f). This cultural hegemony, as Riegel describes it, is reproduced, for example, through the content of the framework curriculum, the composition of classes and the teaching staff, spatial barriers, and the subjective standards of teachers. Misrecognition and incidents of epistemic injustice are virtually a given.

1.2 Institutional: Segregation and hierarchization

Contrary to what the assumption of equal opportunities in schools would suggest, the German school system is based on segregation and hierarchization. What Althusser already made clear in his statements on school as an ideological state apparatus and the associated social order of submission and "falling" (cf. Althusser 1970: 128) is particularly well reflected in the German school system. After a four-year primary education[3], the system differentiates into several forms of secondary schooling.

The basis for the division into different types of schools is the meritocratic model, i.e., division according to a performance principle. "Here, we speak of equal opportunities when 'every person is positioned in society according to their abilities and achievements'" (El-Mafaalani 2020: 61). According to this model, the best-performing students should be in the Gymnasium and the worst-performing students in the Hauptschule[4]. The assessment of who can be classified as better or worse performers is based on the subjective assessments of the respective teachers. These are often influenced by negative attributions to certain social groups. In other words, how students are assessed by their teachers is not independent of the assumptions that teachers have about their students.

An example of this bias is a study by Stefan Hradil at a primary school in Wiesbaden, Germany. There, only 76% of children from the lowest educational and income group with an average grade of 2.0 received a recommendation for the Gymnasium, while children from the highest educational and income group with the same average grade received such a recommendation almost across the board – namely 97% (cf. Bühler-Niederberger 2009). The selection of which children are allowed to attend which types of schools, and thus which educational opportunities are available to them, is therefore not based solely on performance.

---

3 Berlin, Brandenburg, and Mecklenburg-Western Pomerania are the only federal states in Germany that provide six years of primary education. In the other 13 federal states, the starting age for secondary school may vary, but the norm remains four years of primary education.

4 The German school system is divided into several types of secondary schools after four years of elementary school, which offer different educational paths and qualifications. The *Gymnasium* usually leads to university entrance qualification (Abitur) after eight or nine years and prepares students for university studies. The *Realschule* provides a general education and ends with the "Mittlere Reife", which opens the door to vocational training or further education. The *Hauptschule* aims to provide a basic general education and usually ends after nine years with the "Hauptschulabschluss". The *Förderschule* is aimed at students with special needs, for example in the areas of learning, language or physical development. This differentiation usually takes place at an early stage and is controversial, as it strongly determines educational biographies and reinforces social inequalities.

The decisions made by teachers are not necessarily based on explicit racist and/or classist stereotypes that they consciously apply. Instead, they are often unconscious and unintentional attitudes. These attitudes are also referred to as *implicit bias* (cf. Scott 2021: 2). Several studies (cf. Walker and Brigham 2017; Hirn & Scott 2017; Jacoby-Senghor et al. 2016: Scott 2021) show that despite awareness of racism, teachers tend to evaluate the behavior and performance of non-*white* students more negatively. This implicit bias also leads to affected students being suspended more often, for example (cf. Scott 2021). But even without such racist bias-related evaluations, children fare much better when they have parents who can supervise and help with homework and have the time to invest in education; a fact that teachers pick up on and that can result in favoring children from middle class backgrounds.

Another example of implicit bias among teachers can be seen in the comparatively high proportion of migrant students in Förderschulen, special schools for students with learning disabilities. Racialized and migrant students are more often referred to this type of school than those without a migration background.

In addition to the failure of language teaching and distorted perceptions on the part of teachers, the maintenance and design of the school system itself plays a role in the dynamics described by Fereidooni. If not enough students are sent to the Förderschule, the school is threatened with closure. If a school does have to close due to low demand, jobs are at risk. Instead, children who are already marginalized receive inadequate schooling and support in order to prevent school closures. The desperate clinging to outdated school systems once again illustrates that the school system can only function if not all children are given equal opportunities for advancement. In other words, the school system is not only ill-equipped to provide equal opprotunities, it is invested in reproducing inequality. According to this design, the possibility of 'picking up' all learners and taking them along on the path to higher education cannot be considered in reality.

The school system is therefore *based on* segregation and hierarchization. Fereidooni emphasizes that discrimination against schoolchildren who are socially labeled as different is particularly evident in the allocation of different types of schools (cf. Fereidooni 2011: 25). "Schools have a relatively fixed number of places to allocate within an educational hierarchy [...], so that the success of one is always the failure of another" (Bommes & Radtke 1993: 485).

> The reasons for this are not based on the disproportionate need for support of non-German students due to cognitive deficits, but on the lack of language teaching skills in German schools and negative ethnic-cultural attributions on the part of teachers. (Fereidooni 2011: 25)

The division into different types of schools according to, for example, learning types would not be objectionable in itself if different cultural values were not attributed to the types of schools. By dividing schools into different types with different cultural values, learners within these institutions are also portrayed as comparatively worthy of education and social respect. What Nancy Fraser (2000) describes as misrecognition in "Rethinking Recognition" can help to understand the normative implications of the German school system. Because referring to a type of school that does not favor a general university entrance qualification as a final qualification is also a subordination of status. An education at comprehensive, secondary, and vocational schools is often considered less valuable and has a decisive influence on the children's chances of success. If learners are transferred to these schools, a normatively 'successful' future in prestigious and/or well-paid professions becomes more difficult for them. All types of schools, and especially all types of qualifications, are anchored in social value and status: While Gymnasium graduates can consider which university course they want to enroll in after leaving school, graduates of special schools often find themselves in so-called workshops for people with disabilities. There, they are often exploited for an hourly wage of €1.35 and further marginalized by society (cf. Krämer 2021).

Before further exploring the question of whether, how, and where misrecognition and epistemic injustice occur in German schools, we will first examine the reality that stands between schools as institutions and their students: the reality of teachers.

1.3 Personnel: High demands and overworked teachers

> With less than 19 hours of teaching per week, 12 weeks of vacation, and a plethora of bridge days, one could say: I had a part-time job with 60 days of vacation, full pay – which was quite good, by the way – and all the privileges of German civil servant status. Sounds enviable. (El-Mafaalani 2020: 187)

In most cases, however, El-Mafaalani's assumption does not reflect reality. His own experience as a teacher also paints a contrasting picture: in addition to 25 hours of teaching per week, which he carried out in up to three different subjects and 8 to 10 classes, and for which he had to prepare and follow up, he also had to attend numerous conferences, further training, creating 30 school development plans, conducting parent-teacher conferences, planning field trips and class trips, mediating conflicts between students, and much more. There is no question that with this workload, some things have to be neglected.

For El-Mafaalani, this often meant sacrificing lesson preparation and development (cf. El-Mafaalani 2020: 187–9).

> Now you might ask yourself: Where are you supposed to find the time and energy to deal with social inequality in theory and practice? After all, you still have all your other tasks and responsibilities. In addition, legal and curricular changes must be kept up with. It is an enormously multifaceted, complex, and stressful job—even without systematically taking equal opportunity into account. (El-Mafaalani 2020: 189)

El-Mafaalani's descriptions make it clear that addressing social inequalities cannot begin when teachers enter the classroom. By that point, it is already long overdue. When teachers enter the school day unprepared, without any strategies for responding to the complex relationships in the classroom, they are forced to rely on their own knowledge and experience. Of course, it would be desirable if they could draw on a repertoire of independently acquired knowledge and strategies for action, for example from authors of anti-racist education in Germany such as Karim Fereidooni, Maisha-Maureen Auma, Tupoka Ogette, Mohamed Amjahid, Aylin Karabulut, or El-Mafaalani themselves. However, this cannot and should not be a requirement for being able to respond to social inequalities in the classroom. Instead, the social diversity of German classrooms should be an integral part of teacher training.

Another problem in German schools that teacher training cannot prepare for is the lack of staff in the schools themselves. With a shortage of almost 5:000 teachers across Germany (cf. Spiegel 2020), it is not difficult to guess who will end up doing the extra work. What is currently unavailable must be covered by teachers who are already overworked. This redistribution of labor comes at the expense of students—and, of course, the teachers themselves.

If the teaching of the curriculum itself is put on the back burner for a moment, the question arises, particularly from a pedagogical and moral point of view: With such a heavy workload, how are teachers supposed to find the time to give all students, with all their individual characteristics, the recognition they deserve? And how can *all* students be adequately and equally recognized when teachers' perspectives are always distorted by cultural values, assumptions, and socially biased assessment criteria (cf. Weber 2005: 70)?

In the following, two social categories are outlined which can expose students to epistemic injustice and misrecognition due to their membership in marginalized groups.

2. Race and religion as categories of discrimination in schools

In "Mythos Bildung", El-Mafaalani repeatedly argues that class affiliation and parents' level of education are the main factors determining children's educational success in schools (cf. El-Mafaalani 2020: 69). However, a class analysis in the context of German educational institutions is outside the scope of this paper; this focus alone could fill a separate work. In the context of this work, racialization and religion are used as categories of discrimination to illustrate epistemic injustices and misrecognition experienced by students through marginalization and through their status as children and, accordingly, the status group of students.

2.1 Race as a category

At least since the recruitment agreements for guest workers between 1955 and 1973 in West Germany, the treatment of migrants in Germany can no longer be denied: People of color[5] are continuously associated with their migration history and devalued on the basis of it. This conditional right to exist is also evident in the term 'guest workers' itself. Anyone who is a guest in Germany should never feel too comfortable, never unpack their suitcases, never really arrive. The idea that all people living in Germany should have the same rights,

5 In the context of this work, all those people are considered people of color who, due to their own migration histories or those of their families, are associated with group-related pejorative and prejudiced stereotypes and are racialized. In the German context, these may be people from so-called guest worker families or, for example, people with a migration history from Eastern Europe, Asia, Africa, or South America. It should be emphasized that not all people with a migration history in Germany are considered people of color. People who have migrated to Germany from Western European and North American countries, for example, are not stereotyped, degraded, and marginalized according to the same standards.

the same working conditions, the same opportunities for success, and the same recognition still remains a distant dream.

This is also described by author and journalist Mohamed Amjahid in a conversation with Serpil Temiz Unvar, the mother of Ferhat Unvar, who was murdered in the racist attack in Hanau on February 19, 2020:

> I think my parents realized very early on that the German school system—even though they are 'uneducated'—has a certain 'plan' for their children. That migrant children and 'foreign children', so to speak, are automatically supposed to do certain jobs. Which doesn't mean that these jobs are bad. But it's simply absurd that young migrants don't have freedom of choice. That they have to fight against this system. (Amjahid 2021)

Here, Amjahid describes what has already become clear through Fereidooni and El-Mafaalani: Racialized children do not have the same educational and career opportunities as white children; they are not treated with the same respect and consideration. From the outset, they are not regarded as subjects in the German education system, but are stereotyped, generalized, marginalized and degraded. These projections from – mostly dominant – collective hermeneutical resources (Mason 2011) provide a suitable starting point for examining the epistemic and recognition-theoretical level of these degrading experiences of racialized students in Germany. After all, a cardinal point of Fricker's theory of epistemic injustice is that the practice of inferring the credibility of statements and access to knowledge resources based on membership in a socially marginalized group is based on negatively charged and identity-based stereotypes. This raises the question of what negative identity-prejudical stereotypes exist about racialized people in Germany that could impair their epistemic position.

> W: It's not just the language, there are students who simply can't think beyond certain boundaries, right? I'm not judging that now. Mario, for example, you don't notice that he's Croatian, right? Georgios, yes, you can tell he's Greek, and he also has immense problems; he'll have to drop out now. With Boris, you can tell, but that's not very symptomatic. Linda, for example, is a girl who has immense language problems and comprehension problems too; she'll have to drop out as well.
> I: She's Greek, right?
> W: She's Greek. (Weber 2005: 73f)

This excerpt from a conversation among teachers clearly shows how learners' perceived performance deficits are directly linked to their origin and thus justified. Although the teacher even notes that she did not want to 'evaluate' the students 'now' (cf. Weber 2005: 73f), that is exactly what she does. She links characteristics such as intellectual inferiority, comprehension problems and learning difficulties to their origin. For the two teachers, origin and performance are directly related (cf. Helmchen 2019: 85). This is a naturalization of social differences that can clearly be attributed to a racist pattern (cf. Rommelspacher 2011: 29).

Christian Helchen goes on to say that among the 20 most common stereotypes held by white people about people of Turkish origin in Germany are adjectives such as conservative, primitive, emotional, impulsive, and traditional (cf. Helmchen 2019: 85). It is obvious that these distorted representations of migrant and racialized people do not simply bypass school staff. The standards used to measure students' performance, their epistemic position, and their need for recognition are not neutral, but rather the product of one's own socialization, media consumption, teacher training, position within the teaching staff, and—in this case—a lack of confrontation with one's own prejudices. This lack of confrontation is also reflected in the way religious affiliations are treated.

2.2 Religion as a category

According to the Federal Constitutional Court, the state must be a "home for all citizens" – regardless of their religious or ideological beliefs. The state must therefore not identify itself with any particular religious or ideological beliefs. Rather, it must be neutral and tolerant towards all religious and ideological communities. (Federal Ministry of the Interior, Building and Community, n.d.)

What the Federal Constitutional Court has established as the basis for social life in Germany, and thus also for all German institutions, rarely describes the reality of religious minorities. It seems difficult to uphold the claim of neutrality and tolerance towards "all religious and ideological communities" (Federal Ministry of the Interior, Building and Community, n.d.) while people who do not conform to the *white* Christian dominant culture in Germany are continuously physically and verbally discriminated against, marginalized, criminalized, and essentialized (cf. Amjahid 2021). In Germany, Jews and Muslims are the main targets of religious discrimination based on the dominant culture (cf. Willems 2020: 11). Attacks on synagogues (cf. Deutsche Welle 2019) and mosques (cf. Spiegel 2021) are part of everyday life in Germany; wearing religious head coverings such as yarmulkes and hijabs does not guarantee safety from violent attacks—rather, it increases it (cf. Köhler 2019). In schools, at work, and in public life, Jewish and Muslim people are regularly reduced to their religious affiliation, uniformised, questioned as 'experts', denounced by their counterparts' superficial knowledge as having interpretative authority, and regarded as a blank canvas onto which all conflicts that may arise within and between religious communities can be projected (cf. Willems 2020).

Julia Bernstein and Florian Diddens (2020) also describe the discrepancy between promised neutrality and tolerance and the reality of everyday life. For Jewish students, for example, there is hardly any opportunity to develop an identity outside of what sociologist Michal Bodemann calls *memory theater*. Max Czollek writes:

> In memory theater, Jews are important, but as in acting, it is not about them as individuals, but about the role they play—their symbolic significance as representatives of the exterminated, their role as Jews for Germans. (Czollek 2018: 85)

Memory theater does not serve to reflect the diversity of Jewish life in Germany, but rather the "need of German society for reconciliation" (Czollek 2018: 85). As a result, Jewish people are constantly associated with National Socialism, the Shoah, and Israel and reduced to these categories (cf. Bernstein & Diddens 2020). Bernstein and Diddens describe how many teachers have no understanding of anti-Semitism outside of National Socialism, yet insist on claiming interpretive authority over what is anti-Semitic (cf. Bernstein & Diddens 2020: 87).

This epistemic asymmetry (cf. Mason 2011), characterized by a willful hermeneutical ignorance (cf. Pohlhaus 2012)—that is, by holding a deficient or distorted concept of social mechanisms, refusing to revise it, and yet claiming interpretive authority—can also be observed in relation to Islam.

In this context, it is important to mention that religion and race cannot always be considered separately. Although an assumed race does not provide any information about people's religiosity, nor does religion provide any information about their race, the two categories are often mixed together. For example, religion-related stereotypes do not refer exclusively to religion but can also be biologically or culturalist in nature (cf. Willems 2020: 11). With the *racialization of religion*, Iman Attia describes how the attribution of religiosity and the racialization of people are strongly linked (cf. Attia 2019: 138).The historical causality of racialization and religiosity described by Attia also leads to the invisibility of religious and ethnic minorities (cf. Attia 2019: 138). This is because the de-individualization and essentialization of affiliations cause those who do not fit into the expected categories to be forgotten. Ethnic minorities within religious minorities and religious minorities within ethnic minorities remain largely invisible, especially in the context of schools. As a result, the lived realities of Christian Arabs, Black Muslims, Jewish Iranians, or Yazidi Kurds, for example, are not acknowledged in the classroom. The motto seems to be: anyone who is perceived as 'Arab' is automatically assumed to belong to Islam. This insensitive approach to religion and race is particularly problematic in terms of epistemic injustice and misrecognition.

3. Epistemic injustice and misrecognition in schools qua marginalization

Students who are marginalized due to their membership in socially disadvantaged groups, for example because of their race, gender, or religion, are epistemically disadvantaged by negative stereotypes that limit the credibility of their statements. Ben Kotzee writes:

> Epistemic injustice would occur if teachers systematically gave learners less epistemic credit than they deserve due to some negative identity prejudicial stereotype pertaining to learners in a particular setting (or perhaps to children in general). This may be because teachers take learners' testimonies less seriously than they deserve to be taken, or it may be that the dominant culture or the way educational institutions are arranged may privilege 'adult' modes of thinking and talk, leaving children at an unfair epistemic disadvantage. (Kotzee 2017: 326)

Examples of the injustices described by Kotzee are not uncommon in German schools either. In an interview, Sepirl Temiz Unvar reports that she was informed in a personal conversation with the school principal that her son would have no chance of success at the school he had enrolled in (cf. Amjahid 2021). Further descriptions by Unvar in the interview make it clear that racist assumptions form the basis for this preliminary assessment. "Epistemic injustice is done here, because the individual is treated as a typical example of a particular social type, before [he] has been allowed to show who or what [he] is or what [he] does" (Murris 2015: 333). In this example the headmaster portrays the boy as comparatively unworthy of equal educational opportunities on the basis of his identity and reproduces social hierarchies. By denying him respect and esteem, the boy is not recognized as a complete human being. The headmaster fails in his moral and legal obligation as an individual and as a representative of an educational institution to ensure recognition in the form of respect and esteem. "In cases in which we fail to give respect, we not only disrespect another person, but we deny them their personhood; we dehumanize them" (Hänel 2020: 13). As Taylor (1994) points out, the question of respect is not one of politeness. Misrecognition is morally wrong primarily because it harms subjects in their fundamental human abilities.

Marginalized students can also experience epistemic harm and misrecognition on a hermeneutical level. Kotzee writes:

> Teaching a canon of ideas and works by actors from a particular cultural tradition makes that form of culture accessible to students; conversely, not teaching other cultural traditions forecloses students' understanding of that cultural tradition. (Kotzee 2017: 327)

When students are denied access to knowledge that describes their own reality outside the dominant culture, the discrepancy between what happens in their own homes and what is taught in school (especially in terms of perspectives) becomes ever greater. As a result, marginalized students are denied epistemic access to concepts that are relevant to them in the classroom. They feel alienated by knowing and practicing different cultural values from their own homes, and the barrier to sharing these perspectives in class becomes increasingly greater (cf. Messerschmidt 2017; Mangan & Winter 2017).

One example of this is the devaluation of language and language varieties in dominant hermeneutical resources such as schools. Most statements from students that do not correspond to a school or educational language register are devalued and suppressed in the school environment. In addition, languages are categorized as 'good' and 'bad'. While some languages are considered respectable, educated, progressive, and valuable for a resumé, others are suppressed. In Germany, language varieties that contain Arabic or Eastern European terms, for example, and bring language patterns from non-dominant resources into schools are considered deficient, and speakers of these language varieties are associated with negative assumptions (cf. Stokowski

2019: 151). In North America, a similar devaluation of language varieties is taking place with African-American Vernacular English (AAVE), which Kristie Dotson categorizes as an epistemic side effect of colonization (cf. Dotson 2011: 236).

However, marginalized students can be hermeneutically disadvantaged not only in how they speak, but also in what they say. For example, if only limited insights into other cultural traditions are taught in the classroom, students from marginalized groups are often placed in the position of speaking as representatives of an entire culture, an entire country, or a religious group. Suddenly, students find themselves acting as specialists on these topics without necessarily having any actual connection to them (cf. Kotzee 2017: 327). Marginalized students are accorded a surplus of credibility when it comes to talking about their 'own culture'. However, this increase in credibility is conditional: only when it comes to reproducing content that has not already been discussed in class as representatives of an entire group do the students gain this surplus of credibility. Hence, it relates directly to epistemic injustice due to credibility excess (cf. Davis 2011; Lackey 2021; Medina 2011; Yap 2017) and content-based testimonial injustice (cf. Davis 2021; Dembroff & Whitcomb 2023).

Of course, there may actually be an epistemic asymmetry (cf. Mason 2011) here. In other words, it is not unlikely that marginalized students often have a clearer understanding on topics that affect them directly through non-dominant hermeneutical resources such as their own homes, neighborhood centers, circles of friends, or the internet. However, making assumptions about which topics are relevant to which students based on their group membership and expecting students to be available as representatives on these topics at any time deprives them of their subjectivity. Furthermore, as research shows, marginalized knowers are only awarded credibility excess as long as their testimony is in line with the dominant hermeneutical resources and narratives; in other words, students who are singled out as experts on Muslim traditions are awarded credibility excess as long as they describe their religious practices as, for example, traditional, restricting, or patriarchal. Here, too, a misrecognition takes place.

4. Epistemic injustice and misrecognition in schools qua student[6]

If we take another look at the design of German schools as described above, it becomes apparent that instances of misrecognition cannot only occur qua marginalization. Rather, students find themselves in a special position due to their status as students, which encourages misrecognition.

For example, one experience that is not necessarily linked to belonging to a socially marginalized group is that of repeating a year or being placed in a lower school stream. Let us think back to the ideological character of school described by Althusser. The change into a lower grade or school stream can be accompanied by the loss of friends and thus of important relationships of peer recognition, by misrecognition on the part of parents, and by compromised self-recognition (cf. Kammer 2013: 90). When the academic success of some is built on the failure of others and different types of schools vary in their social value (cf. Fereidooni 2011), misrecognition is already inherent in the structure of schools. Transferring students to a different type of school can thus be interpreted as a kind of 'social grading' that clarifies the child's position within society.

The relationship between teachers and learners can also lead to misrecognition. Teachers who accompany students for long periods of time and play a significant role in their school experience are particularly important in recognizing their students on an intellectual and personal level. After all, most children spend the majority of their childhood and youth in educational institutions. Hänel writes: "The misrecognition or lack of recognition from 'significant others' can be especially devastating for our self-recognition" (Hanel 2020: 10).

A lack of institutional recognition by teachers can be caused by framework conditions such as time constraints or overwork. According to Kammler, however, authoritarian behavior, emotional disinterest, a lack of student orientation, a lack of didactic competence, or a lack of democratic values on the part of teachers can also be the basis for recognition deficits (cf. Kammler

6 Although the concept of childism understands children as the target group of group-related injustices, the injustices in the context of this work are named as "qua students". This takes two levels of power into account: those already named by adultism and/or childism and, beyond that, those characterized by the relationship between teachers and learners. Children can take on a special epistemic status as learners, teachers, and intermediaries in the context of school.

2013: 90). Students are particularly prone to diminished self-confidence in their epistemic skills as a result of teachers' and the institutions' misrecognition: First, they are still developing these skills and, thus, orientate on what authority figures such as teachers present to them; second, they spend most of their childhood within schools with few opportunities to receive full recognition or counter misrecognition outside of school; and, third, teachers and school institutions wield the (arbitrary) power of taking students out of peer groups where they can receive recognition as happens in cases where students have to repeat a year or are moved to a different school. Hence, as mentioned above by Hänel, misrecognition can have a direct impact on the self-recognition of students. Misrecognition can be internalized and impaired self-recognition can lead to inferiority issues (cf. Keshky & Samak 2017: 129). Furthermore, if students link their self-worth to capitalist virtues such as productivity or their performance in class, they may also doubt their social value outside of educational institutions due to inadequate grades and experiences of misrecognition.

However, the recognition relationship between teachers and learners is not exclusively characterized by the school-specific aspects outlined here. Even if we could remedy the aspects outlined here, there would still be no status equality between teachers and students. There remains a distinction between adults and children that carries with it a social power imbalance. This is what childism precisely describes: the attitudes and behavior of adults who assume that they are more intelligent, competent, and better than children and young adults solely because of their age (cf. Ritz 2013: 163). According to Elizabeth Young-Bruehl, what is needed above all is the acknowledgment that there are prejudices against children as a distinct target group that undermine their status and make them vulnerable to experiences of violence. She argues, for example, that we can better understand phenomena such as the normalization of physical and psychological violence against children, childhood poverty, child prisons, and child pornography if we understand children as a social group that is subject to prejudice and socially classified as inferior to adults (cf. Young-Bruehl 2012: 6f). Importantly, their biological development is not the sole basis for assumptions about them: children as a group are also subject to social, cultural, and political constructs, evaluations, and distortions about their childhood. The establishment and ubiquity of these assumptions are analogous to those of all other marginalized target groups (cf. Young-Bruehl 2012: 7).

In this sense, we can understand children as a social group that, qua their being children, can become the target of injustice on the basis of preconceptions, beliefs, and resulting actions (cf. Young-Bruehl 2012: 35) – which also form the basis for injustices qua social group in other forms of discrimination[7]. When children are not regarded as full moral persons in comparison to adults because of their being children, when their autonomy is questioned and they are continually classified as non-adults, a clear misrecognition occurs. If adulthood is seen as the goal of all development and adult modes of thinking, speaking, and acting are established as the norm in most contexts, then childhood is devalued as imperfect and incapable (cf. James & Prout 1990: 10f). These instances of misrecognition are evident, for example, in microaggressions that undermine children's value, self-esteem, and dignity through subtle, frequent, and ongoing devaluations. Children are not recognized as complete human beings whose status as children is as legitimate and unconditional as that of adults (cf. Pierce & Allen 1975).

This misrecognition is also found in the context of school. As already mentioned by Ritz and Young-Bruehl, prejudiced stereotypes exist that impair the epistemic status of children qua group membership. Examples of stereotypes about children include that they are irrational, easily influenced, and unreliable (cf. Burroughs & Tollefsen 2016: 366; Brainerd & Reyna 2012: 227). These assumptions can serve as a basis for compromising the credibility of their statements.

If we understand the status of children in the context of adultism (cf. Ritz 2013) as one that assigns them to a social group and, based on this affiliation, exposes them to social, cultural, and political constructs, evaluations, and distortions that can impair the credibility of their statements, the leap to Fricker's concept of testimonial injustice is not far off. In fact, Burroughs and Tollefsen (2016) illustrate testimonial injustice in children with an example from "A Circle of Quiet" (1972), the autobiography of Madeleine L'Engle. She recounts a situation in class that many students are familiar with: When a student needs to

7 The aim of this paper is not to compare the experiences of children as a target group of social injustice with those of racialized people, women, poor people, disabled people, or other socially marginalized groups. It does not claim that children experience oppression similar to victims of racism, sexism, ableism, etc. because they are children. All people who are marginalized and oppressed because of their group membership are subject to specific power structures and specific manifestations and mechanisms of identity-based power (cf. Fricker 2007), all of which must be considered in their own particularities. In addition, children may be exposed to multiple forms of oppression through multiple group memberships, which shape their experiences in all their intersections (cf. Burroughs & Tollefsen 2016).

use the bathroom during class, she asks her teacher for permission three times. Permission is denied three times. By the time the bell rings and the student can go to the bathroom, it is already too late. (A more modern version of the same story is the well-known answer by teachers to the question of whether one can use the toilet: "I don't know if you *can*." This seemingly funny response is funny only at the expense of the student.)

Primarily, the subject is wronged in a testimonial situation in her epistemic capacity as a knower (cf. Fricker 2007: 44). The interaction between the student and her teacher is clearly characterized by identity-based power (cf. Fricker, 28): The teacher is in a position to evaluate the student's statements and declare them invalid. In addition, the situation is shaped by the position of the teacher as an adult and the position of the student as a child (cf. Ritz 2013; Young-Bruehl 2012). In this situation, the identity-based stereotype that children are unreliable and therefore cannot be credible is at work.

When subjects are denied the opportunity to speak for themselves, they can lose self-confidence in their own ideas; this epistemic damage impairs epistemic value and intellectual self-confidence and constitutes a misrecognition (cf. Burroughs & Tollefsen 2016: 375). It is obvious that the testimonial injustice L'Engle experiences here happens qua her status as a student. It is not her gender, sexuality, or class background that impair the credibility of her statements and her epistemic position as a knower in the situation outlined. Instead, the relationship between L'Engle and her teacher, which is influenced by their roles as teacher and student, as well as by their roles as child and adult and the stereotypes associated with these roles, is the basis for the testimonial injustice experienced by the student here. The testimonial injustice thus occurs on the basis of the student's membership in the social group of students.

It is important to note that the teacher's failure to recognize her student as an epistemically capable person and thus also as a person with full moral status is not an individual failure. In addition to the prejudiced stereotypes that play a significant role here, it is also relevant to ask whether dominant ways of speaking, thinking, and acting already perceive adulthood as the norm, mark those of children as deviant, and make it difficult for them not only to articulate their experiences but also to access knowledge resources themselves.

This raises the further question of whether students, due to their status as students, not only experience testimonial injustices, but can also become the target of hermeneutical injustice. There is no clear answer to this question. Michael Baumtrog (2018), for example, argues that the ability to use interpretive resources must first be learned. If learners are unable to use interpretive resources to the same extent as adults due to their less advanced development, the two groups would have unequal but not unjust access to hermeneutical resources (cf. Baumtrog 2018: 303).

However, this outlines the lifeworld and interpretative framework of adults as a goal that must be achieved in order to be recognized as a complete person. The opportunity to view children's lifeworlds as equally complete and legitimate, and to reflect on whether the available access points in the education system actually do justice to these lifeworlds, is missed here. Murris also recognizes the missed potential to actually reflect all content relevant to school learning in the curriculum: "Children's situated lived experiences of learning, their friends, family, or community are irrelevant to the 'real' work in class" (Murris 2013: 248). This discrepancy between potentially learnable and actually taught content can be explained by the fact that adulthood is also understood as the norm in schools (cf. Murris 2013: 253). According to this view, children are hermeneutically marginalized (cf. Fricker 2007: 152); the dominant hermeneutical resource of the school centers on adult language and interpretations of concepts. This suggests that children are only regarded as fully moral persons in social interactions once they have achieved the 'goal' of acquiring adult modes of language and action. The lifeworld of children is thus not considered equal to that of adults (cf. Murris 2013: 257). In the collective hermeneutical resources that Fricker considers as the basis for hermeneutical agency, there is little to no understanding of children. The generic knowers are adults (cf. Baumtrog 2018: 304). The understanding of children thus remains incomplete until they manage to access adult interpretations. The effort to make children's perspectives visible in hermeneutical resources that also contain understandings of children's interpretations remains minimal.

5. Dual epistemic injustice, intersectionality, and epistemic asymmetry: What (can) students know?

From the explanations above, we can conclude that children in schools can experience epistemic injustice and misrecognition both qua their membership of a socially disadvantaged group and qua their status as students. As Murris writes:

> Epistemic injustice is the case with all children (e.g., developmentality). [...] Conceptions of childhood and implicit and naturalized discourses about chil-

> dren inform how adults experience and interpret their educational encounters with children [...]. But when a child is black (and also female), the injustice done to her could be even greater. (Murris 2013: 257)

The various parts of their identity that constitute their position as speakers and knowers must not remain as separate, isolated phenomena. Instead, all parts of identity that influence the subject's position within the framework of standpoint theory should also be linked to each other. Let us think back to Tom Robinson in Lee's *To Kill A Mockingbird (1960)*: Robinson's position as a speaker in the courtroom was not only compromised by his blackness, but by the particular intersection of his blackness and his maleness in the context of the criminalization of black men in the United States (cf. Smiley and Fakunle 2016: 352). Similarly, Carmita Wood was hermeneutically marginalized by several aspects of her identity (cf. Berenstain 2016: 584). Her positions were constituted by *several* parts of her identity; her epistemic position was thus also compromised by several aspects.

5.1 Marginalized and student: a dual injustice

This phenomenon can be better understood through the lens of *intersectionality*. In "A Black Feminist Statement," the Combahee River Collective, a collective of Black lesbian women, made it clear as early as the 1970s that a collective process for the equitable distribution of rights and resources must also consider the diverse perspectives within feminist, anti-racist, and anti-capitalist struggles. To this end, the intersections that may exist within these struggles must also be considered (cf. The Combahee River Collective 2014).

Black women such as Carmita Wood, for example, are situated differently as epistemic subjects than their white female colleagues (cf. Berenstain 2016: 584). Similarly, students such as Ferhat Unvar's brother are situated differently as epistemic subjects than fellow students who conform to the norms of German dominant culture. Baumtrog and Peach (2019) propose that childhood, like other forms of oppression, should also be recognized as an aspect that can impair people's epistemic position. Analogous to ethnocentrism, heterocentrism, and androcentrism, *adultcentrism* can also be seen as a dominant interpretive framework that sets norms and excludes deviations from them (cf. Baumtrog & Peach 2019: 226). Yet, this does not homogenize the experiences of differently situated students; rather, in the intersection of social identities, *being a student* should be considered as yet another identity that plays a role in the diverse experiences of oppression and marginalization.

Accordingly, marginalized students are exposed to a *double* epistemic injustice that can influence their position as speakers and knowers. First, through identity-based prejudicial stereotypes that are applied on the basis of their membership in a socially disadvantaged group, and second, through their membership in the group of children or students, which parallelly expose them to other pejorative identity-based prejudicial stereotypes (cf. Baumtrog & Peach 2019: 227).

The hermeneutic gaps that this creates in the collective dominant resources can be described using Purdie-Vaughn and Eibach's concept of *cultural invisibility* (Purdie-Vaughn & Eibach 2019: 383): The failure to represent the specific experiences of marginalized and intersectionality marginalized groups in cultural memory (cf. Purdie-Vaughs & Eibach 2019: 383).

> As a consequence, 'such people will often be mischaracterized and misunderstood,' as Fricker claims of epistemic injustice in general. [...] This is also problematic for intersectional children. (Baumtrog & Peach 2019: 227f)

As with other forms of epistemic injustice, when children's perspectives are distorted and misrepresented in collective hermeneutical resources, this harms not only the children themselves, who are prevented from making their own experiences understandable to themselves and others, but everyone. In the words of Murris:

> If teachers are not open to the possibility that they can learn from children, then it is a case of 'listening-as-usual' (Davies 2014). It involves a listening out for, or rehearsal of, what teachers already know. Teachers' self-identity as epistemic authorities can constitute a serious barrier to listening to children [...]. (Murris 2015: 334)

If teachers are not open to learning from their students in the classroom and instead imply a supposedly one-sided epistemic asymmetry that always places teachers in a position of epistemic superiority compared to their students, they miss out on the opportunity to benefit from the perspectives of everyone in the classroom. In this context, it is often argued that adults *are* epistemically superior due to their experience and accumulation of knowledge, and that it would be disproportionate to recognize students as equal knowers and grant them the same epistemic authority. An assumption that is deeply distorted.

5.2 Epistemic asymmetry: What can students know?

In feminist pedagogy, and also in this work, it is not at all the intention to discredit the expertise of adults and, in the context of this work, the expertise of teachers. As Carmen Luke puts it:

> Feminist educators, like any academic on the university payroll, *are* institutionally authorized because they are judged and named, at the moment of tenure or hiring, as authorities of knowledge. (Luke 1996: 293)

The question here is not whether teachers should be denied their epistemic status, but whether students can *also* be granted expert status in certain situations.

Based on the arguments presented in this paper, the answer is yes. By granting students epistemic authority, we do not render their status as learners invisible. There is no question that students know less at the beginning of their school career than they do when they graduate from high school. Of course, especially in educational institutions, they are in the midst of a process of finding appropriate terms to make their world understandable to themselves and others. However, this process does not end with adulthood or with the attainment of a university degree. It will probably never be possible to achieve full epistemic authority on a subject and thus be epistemically superior to everyone else. Murris calls the acknowledgment that all knowledge is limited and that a 'complete' mode of knowledge can never be achieved *epistemic modesty* (Murris 2015: 334). Even the oldest members of our society will never be able to complete the process of knowledge production, reproduction, and appropriation. Why, then, is it assumed that they are epistemically superior to children in every regard?

Children can also be experts. For example, when it comes to their own experiences of violence. When children experience violence through domestic abuse or are discriminated against through forms of oppression such as racism, sexism, or anti-Semitism, they often have a better understanding of what is happening to them than someone who can and has remained ignorant of these experiences due to their social position (cf. Burroughs & Tollefsen 2016; Mason 2011; Mills 2007). A teacher who suggests that racism or anti-

Semitism do not exist or do not exist in the form the student is testifying about is likely to have less knowledge about these issues than the students, who often cannot remain ignorant due to their personal circumstances. There is an epistemic asymmetry (cf. Mason 2011) in which the students are epistemically superior to the teacher in this particular context of knowledge acquisition. However, the epistemic position of the students does not stem from the assumptions that teachers make about their students and, in the context of the classroom, do not hesitate to declare them experts on entire countries or religious communities, but is instead based on their own experiences.

Students can assume a similar epistemic position in the context of niche knowledge. For example, if learners have a particular interest in an instrument, marine biology, or astronomy, they may be epistemically better situated on these topics than teachers who have little contact with this subject matter through their school subjects and their own life experiences. Even then, students can act as experts who have acquired a repertoire of knowledge on specific topics through non-dominant hermeneutical resources (cf. Mason 2011) that are hardly or not at all covered by the framework curriculum. The fact that students are younger than their teachers does not necessarily influence their ability to acquire new knowledge. Here, too, there is an epistemic asymmetry in which students may have more knowledge about certain topics than their teachers. Granting this epistemic authority to students does not discredit the epistemic position of the teacher. Ultimately, students are recognized as epistemic subjects who have the ability to acquire knowledge through dominant and non-dominant hermeneutical resources, to manage this knowledge, and to vocalize it for themselves and others. Recognizing that children have the ability to acquire knowledge and bear witness in specific areas does not mean that children are accorded an unreasonable or disproportionate level of credibility and epistemic power. We are simply recognizing them in a fundamental human capacity. In general, it could be said that children's curiosity puts them at an advantage of acquiring knowledge if they are being positioned in a way that allows them to follow their interests. In other words, not only should it be the school's and teacher's responsibility to foster relations of recognition with their students but to create spaces in which children can become knowers.

Interestingly, if we take the idea of recognition seriously, valuing children and students as knowers not only grants recognition to children and students but also to teachers. The main idea of recognition theory is that recognition is necessary for developing a positive relation to self. Recognition is given within a social context of mutual vulnerability. The social practice of giving or bestowing recognition on someone also alters that person in a way that they are placed in a position that allows them to give recognition back. In other words, not fully recognizing another implies that we fail to receive recognition ourselves, since by devaluing another we restrict their capabilities in giving recognition. Mutual recognition between teachers and children or students is thus beneficial for both sides.

6. Principles for a more epistemically just school

So far, it has been shown that children and students are subjected to epistemic injustice not only in cases in which they belong or are taken to belong to a marginalized social group but also qua students. And it has been shown that both forms of epistemic injustice are closely linked to practices of misrecognition that are individually carried out by teachers but also reproduced by the design of school institutions. At this point, the question arises as to what specifically the institution of school can do to counteract the epistemic injustices and instances of misrecognition that are continuously being reproduced there. After all, it is or should be the responsibility of schools to provide all students with equitable access to education. As Kotzee writes: "From a social epistemological perspective, one may say that part of the point of education is to prepare people to take part appropriately, fairly, and *justly* in knowledge exchange" (Kotzee 2017: 329). In the following, I propose some principles for a (more) epistemically just school that bring about revisions of existing structures and norms at both a structural and individual level.

6.1 Breaking old patterns: Inclusive curricula and schools as spaces of solidarity

As already shown earlier, existing framework curricula in schools already constitute a breeding ground for hermeneutical injustice. If the curriculum only reflects certain lived realities, while other students are denied access to knowledge that is relevant to them, some subjects are prevented from conceptualizing their own experiences through collective hermeneutical resources. What is taught is therefore not socially representative. At least not for everyone. They are mainly representative in the sense that they reproduce the construction of the other and the associated deep-rooted social injustices in the classroom content (cf. Fereidooni & Simon 2020).

To counteract the reproduction of this distorted social representation, Kotzee proposes creating a more epistemically just curriculum:

> [...] Achieving hermeneutical justice through education may well involve an attack on elite forms of education that provide the children of some (but not of others) with disproportionate cultural influence in our society and that, at the same time, hold up a certain culture as a pinnacle of education and refinement. (Kotzee 2017: 331)

According to Kotzee, exclusive curricula are problematic primarily because they reproduce the assumption that only selected cultures or certain ways of life – in our case, those of the dominant culture – are epistemically relevant. Students whose real lives differ from those of the dominant culture are thus taught that their cultural values and the ideas and concepts specific to their culture are not relevant to the classroom and therefore not relevant to society. According to Kotzee, preventing this misrecognition requires a more epistemically diverse curriculum (cf. Kotzee 2017: 331).[8]

Yet, whether or not more texts by women, trans people, or people of color are covered in class does not change the fact that students continue to be singled out and degraded because of their group membership. Whether or not there are more teachers of color or more teachers with experiences of racism, sexism, or classism does not change the fact that the school as a system promotes certain lived realities and sanctions the habitus of others. Especially if we continue Althusser's analysis of school as an ideological state apparatus that takes the design of schools to be invested in the reproduction of inequality, a much more fundamental and, above all, structural critique of the institution of school is necessary. Unfortunately, this goes beyond the scope of this project and cannot be pursued further here.

On a *hermeneutical* level, however, epistemic diversity is fundamental. If the aim is to make available in dominant hermeneutical resources all the concepts

8 It should be noted that this argument is not undisputed: Harvey Siegel (2006) questions whether diversifying the framework curricula actually solves the problem. For him, diversity does not necessarily mean cultural diversity. It is sufficient to teach a diversity of research methods and approaches, ideas, and schools in educational institutions. Cultural diversity is not necessarily required (cf. Siegel 2006). It is plausible that epistemic diversity alone cannot create an epistemically just school.

that people need in order to conceptualize their own experiences, then the content of school as a hermeneutical resource must be reworked. If only the lived realities of students from the dominant culture are represented in the curriculum, other students experience an epistemic disadvantage. They are denied the opportunity to acquire the concepts that are relevant to them in the school environment. Kotzee writes:

> […] In order to address hermeneutical injustice, the worst off need to acquire the intellectual tools to make sense of and combat their own oppression; this will be achieved by offering these students the best critical political education that we can muster. (Kotzee 2013: 349)

Bellino and Louckey (2017) go further than Kotzee and argue that it is not only marginalized students who should be given the opportunity to learn about concepts that describe their real lives in the school environment, but all students. With their demand for a school based on solidarity, they ask how we as a community can show solidarity with those who are structurally marginalized by educational institutions while at the same time understanding how these exclusionary mechanisms function in the first place and have gained so much popularity throughout society (cf. Bellino & Louckey 2017: 231f). The point Bellino and Louckey make here ties in with Mills' concept of white ignorance (cf. Mills 2007): It is not exclusively about those personally affected by social injustices to make sense of their experiences of that injustice. It is also important to include all those who can remain ignorant of these mechanisms because they are not targeted by these injustices in their own lives.

In Bellino and Louckey's concept of solidarity, however, it is not one's personal involvement that should be the starting point for action. In a classroom based on solidarity, it is in everyone's interest to understand each other's realities and, above all, to understand the structural mechanisms that constitute these realities. They write:

> It involves people who mobilize ties of mutual understanding and community obligation to confront and change what is unacceptable. Forces of exploitation, capitalization, and dispossession are the main arenas for solidarity, yet its determinants and dynamics also apply to effective and transformative education, including the hidden curricula and implicit hierarchies enacted even in democratic classrooms. (Bellino & Louckey 2017: 229)

Making the curriculum more inclusive and breaking down existing hierarchies in schools is therefore in the interests of all. For it is only in classrooms where social hierarchies based on identity-based power are no longer continuously reproduced that all participants can be equally recognized in their status as knowledgeable individuals. After all, which child contributes which ideas to the classroom is not determined by their group affiliation. In order for children to be recognized for their interests and potential, unlike in the example of Unvar's son (cf. Amjahid 2021), they must first be recognized in their subjectivity. This cannot happen if our perception is distorted by implicit assumptions about social groups. Becoming aware of one's own assumptions is also part of Fricker's virtue of testimonial justice and Medina's epistemic virtues (2012, 2023).

6.2 Creating new spaces: raising awareness, time for recognition, and collective hermeneutical resources

"What sort of critical awareness is needed for a hearer to be able to correct for identity prejudice in a given credibility judgment?" asks Fricker in her chapter on testimonial justice (Fricker 2007: 90). Fricker argues that the listener must be aware of their own social identity and that of their counterpart, including the assumptions that could be made on the basis of these positions. A virtuous listener neutralizes the influence of prejudice on the assessment of the credibility of the other person. Through testimonial sensitivity, listeners can continuously correct their identity-based assumptions and thus approach testimonial justice.

Fricker's virtue of hermeneutical justice also has a corrective character. This means that it is the responsibility of the listener to consider the hermeneutical marginalization of the other person and to take into account any missing or distorted concepts in the speaker's description (cf. Fricker 2007: 169). In a next step, the goal of hermeneutical justice would be to eliminate hermeneutical marginalization altogether and to break down the varying barriers to access collective hermeneutical resources. Yet, Fricker notes that this state cannot be achieved through the individual efforts of virtuous listeners:

> Shifting the unequal relations of power that create the conditions of hermeneutical injustice (namely, hermeneutical marginalization) takes more than virtuous individual conduct of any kind; it takes group political action for social change. (Fricker 2007: 174)

However, Fricker's argument remains primarily at an individual level. Fricker appeals to individuals to become aware of their own social identity and that of others, and to consider, in the long term, to dismantle possible identity-based assumptions that would impair the testimonial and hermeneutical role of speakers. However, individual corrections are not sufficient to address the structural problem of epistemic injustice; if they are at all possible in the context of implicit biases.

Naturally, it should be everyone's concern to correct problematic assumptions about others. This is especially true for teachers, who have a significant influence on how much they value their students. Recognition in the form of respect must be guaranteed by teachers to all learners (cf. Hänel 2020: 14). However, leaving this to the individual responsibility of teachers would be fatal. It should not be a matter of luck whether a student is taught by a teacher who, out of personal interest, is concerned with their social and epistemic position in the classroom and attempts to respect all learners equally.

Raising awareness among teachers must be an integral part of teacher training for everyone. Is there any better place for students to learn how to deal more critically with their own implicit and explicit stereotypes than in a classroom community? Sayles-Hannon writes:

> If [...] the most just testimonial evaluations will occur within an epistemic community, the social justice classroom could be an excellent place to start discussing possible prejudices or systemic ignorance that may influence assessments of testifiers' epistemic authority. (Sayles-Hannon 2012: 387)

This requires the teacher to not only engage critically with the task and responsibility of being a responsible listener for the students, supporting them in their access to hermeneutical resources and critically examining what is considered legitimate knowledge in society and what is not, but also to address the multitude of discriminations that are continuously reproduced through teaching.

Finally, awareness of racism and sexism, anti-Semitism, anti-Muslim racism, and other forms of oppression is necessary in order to be able to identify racist, anti-Semitic, or sexist images in teaching content and classroom discussions in the first place. That is why Fereidooni and Simon (2020) demand:

> In a migrant society, anti-racist subject-specific teaching methods are an absolute necessity for (future) teachers, as (only) a deconstruction of racism-related issues within subject-specific teaching methods can prevent racism from being (unquestioningly) reproduced. In order to achieve anti-racist lesson planning and thus anti-racist subject-specific teaching considerations, subject-specific phenomena must therefore be reflected upon and deconstructed from an anti-racist perspective. (Fereidooni & Simon 2020: 1)

Such a critical reflection is, of course, not limited to anti-racist teaching methods, but can be extended to other forms of oppression, including adultism (cf. Ritz 2013). In contrast to racism, for example, adultism and childism are hardly ever considered in the context of schools. Yet, adultism significantly shapes the relationship between teachers and learners, including their status as knowledgeable individuals. Teachers can have a clear awareness of racist assumptions and actively work to prevent them from distorting their view of their racialized students. Despite such efforts on the part of teachers, adultism nevertheless deeply affects both their status as knowledgeable individuals and the personal status of their students. Hence, teachers might still engage in practices of misrecognition of students qua their status as students.

Furthermore, as long as the design of school institutions prevents teachers from engaging in epistemically just and recognition-based practices, individual virtues and anti bias training will not do the trick. As El-Mafaalani (see 2020: 187–9) already pointed out, most teachers are already overwhelmed by the demands placed on them as teachers. Overcrowded classrooms and a number of teaching hours that for many is incompatible with a regular 40-hour week mean that there is not enough time for sufficient mutual relationships of recognition between teachers and students (cf. El-Mafaalani 2020: 60). Plus, when teachers are under time constraints, they are more likely to rely on implicit biases (cf. Berliner 1990; Scott 2021). Hence, institutions have to provide work contexts in which teachers can reduce their workload. It is only when teachers are no longer forced to resort to implicit biases under time pressure in order to make decisions more quickly that they can start to recognize their students sufficiently and independent of their social position.

Finally, more needs to be done to promote these ideas and projects. Carel (2014), Györffy and Murris (2013, 2015) as well as Burroughs and Tollefsen (2016) have already shown that some of the forms of epistemic injustice experienced by children stems from the centering of adult modes of thinking and speaking. This results both in testimonial injustices and in problematic practices on a hermeneutical level insofar as there are insufficient collective hermeneutical resources in which children are granted epistemic power and can create and manage hermeneutical resources themselves. Schools could be the ideal space for fostering collective hermeneutical resources. After all, it is in the context of schools where students learn how to critically engage with resources of knowledge, use different sources to acquire clear concepts, and record these in various forms. For example, students could work collectively on glossaries containing terms that are relevant to them and defined by them, instead of having to rely on existing ones, which may contain interpretations that differ from their own experiences. In fact, giving children opportunities to produce knowledge does not prevent them from using existing hermeneutical resources and acquiring knowledge from them, rather, it could be a valuable addition to their school education, through which they learn to work in a self-determined and self-organized manner. Self-determination and self-organization are fundamental skills for feminist education and are in any case neglected in schools (cf. De Santis & Serafini 2015).

To further strengthen these skills, Haynes (2009) suggests involving students more closely in school structures and assigning them roles that allow them to have a say in their everyday school life.

> The 2004 Ofsted report indicates that a key ingredient in the school's success is that children know they are listened to and that their contribution to the development of the school community is valued. (Haynes 2009: 59)

Haynes makes it clear that shared management enables learners to better represent their concerns and gives them opportunities to change things that have previously limited rather than promoted the progress of their skills at school. Haynes suggests, for example, that students be given a sum of money to enable them to realize their own projects and make the school their own space (cf. Haynes 2009: 60). "Children are not citizens-in-waiting, they already belong to communities and can participate in ways that benefit those communities" (Haynes 2009: 61). It is the school's responsibility to recognize students in their irreplaceable role and show them how fundamental their contribution is to the school community.

6.3 Recognizing (in)equalities: epistemic (a)symmetry, epistemic dependence, and epistemic modesty

Part of this task is also to acknowledge that education is not a process in which knowledge is passed on linearly from teachers to students. Instead, the focus should be on the process of teachers and learners finding meaning and understanding together in the classroom. There is a kind of *epistemic dependency* in which teachers and learners are mutually dependent on each other for the knowledge that emerges in the classroom. Kotzee writes:

> Rather than conceiving education as a process by which knowledge is passed on from teacher to learner, Murris [...] insist that one must see teacher and learner as making meaning together in the classroom. (Kotzee 2017: 328)

Teachers depend on ideas, established connections, and the learning process of their students. Otherwise, teaching cannot take place. Students depend on the expertise and guidance of their teachers; otherwise, they cannot contribute to developing meaning and significance within the classroom community.

While such epistemic dependency seems obvious, it is often not made visible, let alone acknowledged. This is probably due to a fear that acknowledging epistemic dependence could undermine authority and expertise. Yet, according to theories in feminist pedagogy (cf. Luke 1996: 293), recognizing epistemic dependence simply means that students are also knowledgeable subjects who can contribute significantly to the progress of the lesson content.

> In the longer term, one may hold that the very point of education is to make it the case that learners can begin to assign the right amount of epistemic authority to speakers generally and that, in those cases in which they are themselves the epistemic authorities, they assume this responsibility and discharge it well. From a social epistemological perspective, one may say that part of the point of education is to prepare people to take part appropriately, fairly, and justly in knowledge exchange. (Kotzee 2017: 329)

Kotzee makes it clear here that part of the competence to attribute an appropriate degree of credibility to speakers also includes being able to act with epistemic authority oneself. Learners may have epistemic authority over their own social experiences or their acquired niche knowledge. There is therefore necessarily an epistemic (a)symmetry. On the one hand, the potential of students and teachers to act as learners and educators simultaneously should be recognized as symmetrical. On the other hand, it is important to recognize the expertise of teachers and students where it exists. In order to recognize this asymmetry from the perspective of students, teachers must remain open to learning from them as well. Murris writes:

> When thinking with children, adults need to 'give' their mind to what there is to think about, which is only possible when adults are 'open-minded', have 'epistemic modesty', 'epistemic trust' and are committed to 'epistemic equality'. (Murris 2013: 258)

This takes us full circle: While schools need to provide contexts in which mutual learning can be realized, individuals have to train epistemic modesty (Murris 2015: 334), recognizing that all knowledge is limited and that a 'perfect' mode of knowledge can never be achieved, neither by adults nor by children. After all, knowledge evolves with the world it seeks to describe. Remaining open to revising and expanding knowledge already acquired through the experiences of others neither delegitimizes what already exists nor grants speakers an unreasonable or disproportionate degree of credibility and epistemic power. Instead, we recognize that we can keep learning; from each other, with each other, and above all, *through* each other. And we owe this recognition to everyone—including children.

7. Conclusion

This marks the end of this work. However, the discussion on epistemic injustice affecting children is by no means over. There is much more to be discussed and integrated into the discourse on epistemic injustice and theories of recognition. For example, an investigation of parallels between the epistemic status of children in court testimony situations (cf. Oates 2007), a closer look at the implicit bias in the assessment of the performance and behavior of marginalized students (cf. Walker and Brigham 2017; Hirn & Scott 2017; Jacoby-Senghor et al. 2016: Scott 2021) or the epistemic injustice experienced by trans children when they try to make their own experiences understandable to others (cf. Fricker & Jenkins 2017) could be the basis of other noteworthy discussions.

The aim of this project was to illustrate that children in general and students in the context of school can be and are targets of testimonial and hermeneutical injustice, best understood as a form of misrecognition. Yet, the observation of German schools has made it possible to show that children who are already marginalized due to their social group membership are facing multiple epistemic injustices: They experience epistemic injustice qua their social group membership *and* qua being students (or children). Furthermore, this production of inequality is reproduced primarily not be teachers but by the design of school institutions: Misrecognition, thus, occurs through the school system itself.

References

Althusser, Louis (1970): "Ideologie und ideologische Staatsapparate. Anmerkungen für eine Untersuchung", in: La Pensée 151, S. 108–153.

Amjahid, Mohamed (2021): "Talk mit @seerpiltemiz_unvar", May 14, 2021. Instagram Live-Video, 46:59 Minuten. Last used: 15. Mai 2021. https://www.instagram.com/p/CO2wnWZKMq-/.

Arani, Aliyeh Yegane (2020): "Die multireligiöse Schule als Ort der Diskriminierung", in: Joachim Willems (Hg.), Religion in der Schule. Pädagogische Praxis zwischen Diskriminierung und Anerkennung, Bielefeld: transcript, S.163-190.

Attia, Iman (2019): "Unzumutbare Koexistenz: Rassialisierungsprozesse von Muslimen und Musliminnen in historischer Perspektive", in: Wassilis Kassis/Bülent Uçar (Hg.), Antimuslimischer Rassismus und Islamfeindlichkeit, Osnabrück: V&R unipress, S. 125–140.

Baumtrog, Michael D. (2018): "Navigating a Necessary Inequality: Children and Knowledge- Based Injustice" in: Social Inequality and the Spectre of Social Justice. Alternate Routes: A Journal of Critical Social Research 29, S. 294–306.

Baumtrog, Michael D./Peach, Harmony (2019): "They can't be believed: Children, intersectionality and epistemic injustice", in: Journal of Global Ethics 15 (3), S. 213–232.

Bellino, Michelle/Louckey, James (2017): "Education as Solidarity", in: Anthropology & Education Quarterly 28, S. 229–232.

Berenstain, Nora (2016): "Epistemic Exploitation", in: Ergo 22 (3), S. 569–590.

Berliner, David C. (1990): "What's All The Fuss About Instructional Time?", in: The Nature of Time in Schools. Theoretical Concepts, Practitioner Perceptions. New York London: Teachers College Press.

Bernstein, Julia/Diddens, Florian/Friedlender, Nathalie/Theiss, Ricarda (2018): ‚Mach mal keine Judenaktion!‘ Herausforderungen und Lösungsansätze in der professionellen Bildungs- und Sozialarbeit gegen Antisemitismus, Frankfurt: Frankfurt University of Applied Sciences.
Bernstein, Julia/Diddens, Florian (2020): ‚"Traut euch, steht auf, wer hier Jude ist!: Antisemitismus an Schulen. Eine Rekonstruktion der Perspektiven betroffener Schüler*innen und Lehrer*innen", in: Joachim Willens (Hg.), Religion in der Schule. Pädagogische Praxis zwischen Diskriminierung und Anerkennung, Bielefeld: transcript, S. 77–104.
Bommes, Michael/Radtke, Frank-Olaf (1993): "Institutionelle Diskriminierung von Migrantenkindern. Die Herstellung ethnischer Differenz in der Schule", in: Zeitschrift für Pädagogik 39 (3), S. 483–491.
Brandmayr, Michael (2017): "Lernen und die gesellschaftliche Reproduktion: Vier Thesen über die Wirksamkeit von Ideologie in der Schule", in: Zeitschrift für Sozialen Fortschritt 6 (3), S. 187–202.
Bundesministerium des Inneren, für Bau und Heimat: Religionsverfassungsrecht. o.J. Last used May 17, 2021. https://www.bmi.bund.de/DE/themen/heimat-integration/staat- und-religion/religionsverfassungsrecht/religionsverfassungsrecht-node.html
Burroughs, Michael D./Tollefsen, Deborah (2016): "Learning to Listen: Epistemic Injustice and the Child", in: Episteme 13 (3), S. 359–377.
Brainerd, C. J./Reyna, V. F. (2012): "Reliability of Children's Testimony in the Era of Developmental Reversals", in: Developmental Review 32, S. 224–67.
Brühlmeier, Arthur (1980): "Die Noten als pädagogisches Problem: Grundsätzliche Erwägungen zur versuchsweisen Änderung der Notenpraxis im Lehrerseminar St. Michael Zug", in: Schweizer Schule 10.
Carel, Havi/Györffy, Gita (2014): "The art of medicine: Seen but not heard: Children and epistemic injustice", in: The Lancet 384, S. 1256–1257.
Carel, Havi/Kidd, Ian James (2014): "Epistemic injustice in healthcare: A philosophical analysis", in: Medicine, health care, and philosophy 17 (4), S. 529–540.
Czollek, Max (2018): "Keine Juden mehr für Deutsche?", in: Walter Homolka/Jonas Fegert/Jo Frank, "Weil ich hier leben will…". Jüdische Stimmen zur Zukunft Deutschlands und Europas, Freiburg: Herder, S. 84–99.
Davis, Emmalon (2016): "Typecasts, Tokens, ans Spokespersons: A Case for Credibility Excess as Testimonial Injustice", in: Hypatia 31 (3), S. 485–501.

Davis, Emmalon (2021): "A Tale of Two Injustices: Epistemic Injustice in Philisophy", in: Jennifer Lackey (Hg.), Applied Epistemology, New York: Oxford University Press, S. 215–250.

Dembroff, Robin/Whitcomb, Dennis (2023): "Content Focused Epistemic Injustice", in: Oxford Studies in Epistemology 7, S. 48–70.

Dotson, Kristie (2011): "Tracking Epistemic Violence, Tracking Practices of Silencing", in: Hypatia 26 (2), S. 236–257.

Du Bois, W.E.B. (1897): "Strivings of the Negro People", in: The Atlantic August 1897 Issue.

El-Mafaalani, Aladin (2020): Mythos Bildung. Die ungerechte Gesellschaft, ihr Bildungssystem und seine Zukunft, Köln: Kiepenheuer & Witsch.

El-Tayeb, Fatima (2016): Undeutsch. Die Konstruktion des Anderen in der postmigrantischen Gesellschaft, Bielefeld: transcript.

Fereidooni, Karim (2011): Schule – Migration – Diskriminierung. Ursachen der Benachteiligung von Kindern mit Migrationshintergrund im deutschen Schulwesen, Wiesbaden: VS Research.

Fereidooni, Karim/Simon, Nina (2020): "Rassismus(kritik) und Fachdidaktiken – (K)ein Zusammenhang? – Einleitende Gedanken", in: Karim Fereidooni/Nina Simon (Hg.), Rassismuskritische Fachdidaktiken. Theoretische Reflexionen und fachdidaktische Entwürfe rassismuskritischer Unterrichtsplanung, Wiesbaden: Springer VS.

Fraser, Nancy (2000): "Rethinking Recognition", in: New Left Review 3, S. 107–120.

Fricker, Miranda (2007): Epistemic Injustice: Power and the Ethics of Knowing, Oxford: Oxford University Press.

Giladi, Paul (2018): "Epistemic Injustice: A role for recognition?", in: Philosophy and Social Criticism 44 (2), S. 141–158.

Hänel, Hilkje (2020): "Hermeneutical Injustice, (Self-)Recognition, and Academia", in: Hypatia 35 (2), S. 1–19.

Haynes, Joanna (2009): "Listening to the voice of child in education", in: Suanne Gibson/Joanna Haynes (Hg.), Perspectives on participation and inclusion: Engaging education, London: Continuum, S. 27–43.

Helmchen, Christian (2019): Stereotype Threat im Englischunterricht: Zu den Bildungschancen von Jugendlichen mit Migrationshintergrund, Wiesbaden: Springer VS.

Hirn, Regina G./Scott, Terrance M (2014): "Descriptive analysis of teacher instructional practices and student engagement among adolescents with and without challenging behavior.", in: Education and Treatment of Children 36 (4), S. 585–607.

Hirose, Akihiko (2011): "‚No Asians working here': racialized otherness and authenticity in gastronomical Orientalism", in: Ethnic and racial studies 34 (9), S. 1482–1501.

Honneth, Axel (1992): Kampf um Anerkennung. Zur moralischen Grammatik sozialer Konflikte, Frankfurt am Main: Suhrkamp.

Honneth, Axel (2007): "Recognition as Ideology", in: Bert van den Brink/David Owen (Hg.), Recognition and Power: Axel Honneth and the Tradition of Critical Social Theory, Cambridge: Cambridge University Press, S. 323–347.

Honneth, Axel (2011): Das Recht der Freiheit: Grundriss einer Demokratischen Sittlichkeit, Berlin: Suhrkamp.

James, Allison/Prout, Alan (1990): Constructing and Reconstructing Childhood: New directions in the sociological study of childhood, Oxford: Routledge.

Karabulut, Aylin (2020): Rassismuserfahrungen von Schüler*innen: Institutionelle Grenzziehungen an Schulen, Wiesbaden: Springer VS.

Kammler, Tilmann (2013): Anerkennung und Gewalt an Schulen. Eine evidenzbasierte und theoriegeleitete Interventionsstudie im Praxistest, Wiesbaden: Springer VS.

Keshky, Mogeda El Sayed El/Samak, Yasser Abdelazim Abdelmangoud (2017): "The development of self-esteem in children: Systematic review and meta analysis", in: International Journal of Psychology and Behavior Analysis 3, S. 128.

Kotzee, Ben (2013): "Educational Justice, Epistemic Justice, and Leveling Down", in: Educational Theory 63 (4), S. 331–349.

Kotzee, Ben (2017): "Education and Epistemic Injustice", in: Ian James Kidd/José Medina/Gaile Jr. Pohlhaus (Hg.), The Routledge Handbook of epistemic injustice, New York: Routledge Handbooks, S. 324–335.

Lackey, Jennifer (2021): "Eyewitness testimony and epistemic agency", in: Noûs 56 (3), S. 696–715.

Lee, Harper (1960): To Kill a Mockingbird, London: William Heinemann.

L'Engle, Madeleine (1972): A Circle of Quiet, New York: Farrar, Straus und Giroux.

Luke, Carmen (1996): "Feminist Pedagogy Theory: Reflections on Power and Authority", in: Educational Theory 46 (3), S. 283–302.

Mangan, Doireann/Winter, Laura Anne (2017): "(In)Validation and (mis)recognition in higher education: the experiences of students from refugee

backgrounds", in: International Journal of Lifelong Education 36 (4), S. 486–502.
Mason, Rebecca (2011): "Two Kinds of Unknowing." In: Hypatia 26 (2), S. 294–307.
Medina, José (2011): "The Relevance of Credibility Excess in a Proportional View of Epistemic Injustice: Differential Epistemic Authority and the Social Imaginary", in: Social Epistemology 25 (1), S. 15–35.
Medina, José (2012): The Epistemology of Resistance: Gender and Racial Opression, Epistemic Injustice, and the Social Imagination, New York: Oxford University Press.
Medina, José (2023): The Epistemology of Protest: Silencing, Epistemic Activism, and the Communicative Life of Resistance, New York: Oxford University Press.
Messerschmidt, Anne (2017): "Komplexität annehmen – Verflechtungen von Sexismus und Rassismus reflektieren gegen einen migrationsfeindlichen Konsens", in: Johanna Bröse/Stefan Faas/Barbara Stauber (Hg.), Flucht, Wiesbaden: Springer Fachmedien, S. 21–35.
Mills, Charles (2007): "White Ignorance", in: Shannon Sullivan/Nancy Tuana (Hg.), Race and Epistemologies of Ignorance, Ithaca: Suny Press, S. 13–38.
Murris, Karin (2013): "The Epistemic Challenge of Hearing a Child's Voice", in: Studies in Philosophy and Education 32, S. 245–259.
Murris, Karin (2015): "Listening-as-Usual: A Response to Michael Hand", in: Studies in Philosophy and Education 34, S. 331–335.
Oates, Kim (2007): "Can We Believe What Children Tell Us?", in: Journal of Paediatrics and Child Health 43, S. 843–847.
Pierce, Chester M./Allen, Gail B. (1975): "Childism", in: Psychiatric Annals 5 (7), S. 15–24.
Pohlhaus, Gaile (2012): "Relational Knowing and Epistemic Injustice. Toward a Theory of Willful Hermeneutical Ignorance", in: Hypatia 27 (4), S. 715–735.
Purdie-Vaughns, Valerie/Eibach, Richard P (2008): "Intersectional Invisibility: The Distinctive Advantages and Disadvantages of Multiple Subordinate-Group Identities", in: Sex Roles 59 (5–6), S. 377–391.
Riegel, Christine (2016): Bildung – Intersektionalität – Othering. Pädagogisches Handeln in widersprüchlichen Verhältnissen, Bielefeld: transcript.
Ritz, ManuEla (2013): "Adultismus – (un)bekanntes Phänomen", in: Petra Wagner (Hg.), Handbuch Inklusion. Grundlagen vorurteilsbewusster Bildung und Erziehung, Freiburg: Herder, S. 163–173.

Rommelspacher, Birgit (2011): "Was ist eigentlich Rassismus?", in: C. Melter/P. Mecheril (Hg.), Rassismuskritik: 1. Rassismustheorie und -forschung, Schwalbach am Taunus: Wochenschau Verlag, S. 25–38.

Sayles-Hannon, Sally (2012): "On Whose Authority? Issues of Epistemic Authority and Injustice in the Social Justice Classroom", in: Randall Curren (Hg.), Philosophy of Education, New York: Wiley Blackwell, S. 380–388.

Scott, Terrance M (2021): "Implicit bias, disproportionate discipline, and teacher responsibility for instruction as prevention", in: Preventing School Failure: Alternative Education for Children and Youth 65 (4).

Siegel, Harvey (2006): "Epistemological Diversity and Education Research: Much Ado about Nothing Much?", in: Educational Researcher 35 (2), S. 3–12.

Smiley, Calvin John/Fakunle, David (2016): "From 'brute' to 'thug': The demonization and criminalization of unarmed Black male victims in America", in: Journal of Human Behavior in the Social Environment 26 (3–4), S. 350–366.

Stokowski, Margarete (2019): "Sprache", in: Fatma Aydemir/Hengameh Yaghoobifarah (Hg.), Eure Heimat ist unser Albtraum, Berlin: Ullstein, S. 150–155.

Taylor, Charles (1994): "Politics of Recognition", in: Multiculturalism. Examining the Politics of Recognition, Princeton: Princeton University Press.

The Combahee River Collective (2014): "A Black Feminist Statement", in: Women's Studies Quarterly 42 (3/4), S. 271–280.

Weber, Martina (2005): "Ali Gymnasium – Soziale Differenzen von SchülerInnen aus der Perspektive von Lehrkräften", in: Franz Hamburger/Tarek Badawia/Merle Hummrich (Hg.), Migration und Bildung. Über das Verhältnis von Anerkennung und Zumutung in der Bildungsgesellschaft, Wiesbaden: Springer VS, S. 69–79.

Willems, Joachim (2020): "Einleitung", in: Joachim Willens (Hg.), Religion in der Schule. Pädagogische Praxis zwischen Diskriminierung und Anerkennung, Bielefeld: transcript.

Yap, Audrey S. (2017): "Credibility Excess and the Social Imaginary in Cases of Sexual Assault", in: Feminist Philosophy Quarterly 3 (4), S. 1–24.

Young-Bruehl, Elisabeth (2012): Childism. Confronting Prejudice Against Children, New Haven London: Yale University Press.

Musings

Recognition as an Ethic of Living Beings
A Brief Proposal

Karen Ng

Can the concept of recognition provide a meaningful framework for a distinctly *ethical* understanding of the relationship between humans and non-human nature? There are some immediate barriers to answering this question in the affirmative. In both historical and contemporary writings on the topic, the concept of recognition is employed to articulate a unique relation and interaction between two individuals who are potentially or actually equals. The equality in question concerns their status as deserving of certain forms of respect and treatment as free and rational persons, and the adjectives, *mutual* or *reciprocal*, that are so often attached to the concept, are inseparable from the very idea. By definition then, recognition applies exclusively to a relation between human beings, and even more emphatically, according to Fichte, between individuals with a recognizable "human shape" (see FNR §§5-7).[1] In the more well-known, Hegelian paradigm of recognition, this idea is further developed into a theory of historically developing social practices and institutions that are essential for mediating mutual recognition — in the shapes of love, respect, and esteem — in complex human societies. At first glance, the prospects for employing the concept of recognition for understanding relationships between human beings and non-human nature do not look at all promising.

Acknowledging that there is something of an uphill battle to employing recognition as a viable concept beyond the scope of human relationships, I will nonetheless attempt to develop a brief proposal that this concept can provide a helpful framework for an ethical understanding of relations within living nature. I use the term *ethical* here in a broad sense recalling Hegel's conception of ethical life or *Sittlichkeit*, which he also refers to as "the living good [*das lebendige Gute*]" (PR §142). Ethical relationships draw on character, dispositions, and the

1 All abbreviations are listed in references below.

cultivation of habituated interactions, which gain their meaning from species-specific modes of living and can be judged as good or bad for some aim. For self-conscious creatures, ethical life and relationships can give rise to expectations or obligations, but the latter are not necessary for something to have an ethical character. My suggestion will be that we can understand the relations between living beings as ethical in this broad sense, and that recognition is an appropriate concept for capturing, at least in part, the ethical character of these relationships. There are two key aspects to expanding the concept of recognition that I will pursue here. First, it is important to reestablish the essential connection between recognition and the concepts of life and love, which is most clearly articulated in Hegel's early writings. Recognition, paradigmatically expressed in love and friendship,[2] crucially involves the apprehension of life and living form. Second, I will consider the possibility of proto-recognitive activities and processes in non-human life-forms, as well as the possibility of recognition and proto-recognition across life-forms, both human and non-human. In exploring these possibilities, what will hopefully emerge is the importance of the concept of recognition for understanding non-instrumental and ethical relationships among and between living beings, expanding the concept beyond its current, human-centric application.[3]

***

2 "We already possess this freedom in the form of feeling [*Empfindung*], for example in friendship and love. Here, we are not one-sidedly within ourselves, but willingly limit ourselves with reference to an other, even while knowing ourselves in this limitation as ourselves" (PR §7Addition).

3 Several recent works try to extend recognition as an ethical concept beyond the human domain. In a forthcoming volume on the concept of recognition edited by Matthew Congdon and Thomas Khurana, Jay Bernstein argues for the idea of the recognition of nature's right through the lens of ecocide, and Owen Ware argues for the recognition of non-human animals as persons through classical Yoga philosophy. Heikki Ikäheimo does not directly discuss the recognition of non-human animals, but leaves open "the question of whether animals of other species could, in principle, develop person-making features" (2022: 69). Chapter three of Connie Wang's dissertation on Hegel's Anthropology (2025) argues for a Hegelian conception of disregard for the natural world, akin to disrespect and misrecognition, that is indicative of a kind of psychic illness or *Verrücktheit*. Although here I am interested in expanding the concept of recognition beyond the human domain, elsewhere I have defended a conception of humanism that attends to species-specificity without speciesism, one that is compatible with inter-species recognition (see Ng 2021).

Let me begin with two brief passages from Hegel's fragment on love, written in late 1797 or early 1798:

> True union, or love proper, exists only between living beings who are alike in power and thus in one another's eyes living beings from every point of view; in no respect is either dead for the other. (FL 1:245-246/304)[4]
> ...love is a feeling [*Gefühl*] of something living... (FL 1:246/305)[5]

What exactly is the conceptual connection between love and life suggested by Hegel here? Tracing the development of his philosophical system, Dieter Henrich has argued that love and life share two key conceptual characteristics that are central for the basic operation of Hegelian thought. First, the movements of love and life exhibit self-reference and autonomy; second, their movements involve separation and opposition, such that self-reference and autonomy are attained precisely through a relationship to an 'other' (Henrich 2003: 314). While helpful, this is admittedly rather abstract, and Henrich is interested in tracing a thread from Hegel's early writings to his mature *Science of Logic*, demonstrating how he arrives ultimately at a logic of negation as the basis of his philosophy. Moving in a different direction, I want instead to consider the more substantive connection that Hegel draws between love and life in these passages, which allows us to understand some distinctive features of recognition.

To start, Hegel suggests that love is a relation and feeling that can only exist between living beings. Why is this the case and how should we understand this claim? On the face of it, there is something obvious about the claim: insofar as love consists (at least) of feeling, and feeling is something that only obtains for living beings, love is something that can only be felt by living beings. But the claim says more than this: love exists *between* living beings who are alike in power. This builds reciprocity into the feeling of love, such that proper or true love requires some degree of what Hegel will later call a "doubled movement" in his famous account of recognition in the *Phenomenology of Spirit* (PhG 3:146). Love is a reciprocal feeling in which each feels toward the other, and this feeling toward the other is at the same time, for both, an enactment of a feeling of self. That there can be a degree of reciprocal feeling and self-feeling is what

4 "Wahre Vereinigung, eigentliche Liebe findet nur unter Lebendigen statt, die an Macht sich Gleich und also durchaus füreinander Lebendige, von keiner Seite gegeneinander Tote sind." See Katz 2020 who takes up this quotation to develop an understanding of marital love and sexual hierarchy in Hegel's *Philosophy of Right*.

5 "...die Liebe ein Gefühl des Lebendigen ist..."

makes the living beings in question "alike in power." Crucially, love accommodates degrees and modalities of reciprocity without requiring the demanding sense of political or moral equality essential for full-fledged respect recognition. In this context, what matters for being alike in power concerns their character and mutual relation simply as living beings. Elaborating on this feeling between living beings who are alike in power, Hegel suggests that they are for each other "thoroughly [*durchaus*] alive," and in no way dead.[6] Love is thus, fundamentally, a feeling and apprehension that grasps living form and a relation to another insofar as this other is thoroughly alive — that is, dynamic, in development, purposive, self-producing, corporeal, needy, desirous, finite, and vulnerable.

The substantive connection between love and life concerns the importance of the distinction between life and death for the feeling of love. This, I will argue, extends into the concept of recognition, although not in the usual way in which this is understood, where the focus is on the importance of a life and death struggle, rather than on the connection between recognition and the apprehension of living form and the kinds of agency this engenders. Living form and the distinction between life and death matters for love for at least two reasons. First, the distinction helps to mark the difference between relating to an object or mere thing and relating to a living object or *subject-object*. This difference, I will argue, is the difference that makes intelligible and possible the relation of recognition, but for now I want to focus on the broad form of agency that is opened up in relating to a living subject-object as opposed to a mere object. In the reciprocal feeling of living form, one grasps in the other a form of activity that is self-organizing and self-determining—the feeling for the other as thoroughly alive grasps that this other sets its own ends. In the language of Kant's *Critique of the Power of Judgment*, the living being is a *natural end*, or a "cause and effect of itself" (CJ §64). Insofar as this other being sets its own ends (consciously or unconsciously, with intention or inchoately), it can be distinguished from other objects that do not, from things that are mere means without their own self-determined ends. In relating to a living subject-object, what is opened up is the possibility of a relation to another being not as a mere means, but as a being that sets its own ends and that our action might be shaped to take those ends into account. This mirrors a relation that is found in the feeling of respect,

6 Knox's translation is a creative interpretation, but I think the idea that the beloved is grasped as living "from every point of view" helpfully captures what is distinctive about a loving relation to another.

but far from being restricted to rational beings who can represent the moral law, Hegel presents this structure as first possible in the reciprocal feeling of living form — the reciprocal feeling of beings who set their own ends as the cause and effects of themselves.[7] Broadly, then, what the feeling for the other as thoroughly alive opens up is the possibility of a non-instrumental mode of relationship and non-instrumental action in relation to living others. Love as a relation to another being as thoroughly alive, as purposive, dynamic, end-setting activity, enables the possibility of non-instrumental agency and interaction between living beings.

The distinction between life and death matters for love for a second reason: in connecting love so closely with life, Hegel appears to be suggesting that love can only be felt for living beings and not for mere objects. How then can we account for what we feel as a love for places, objects of sentimental, historical, or aesthetic value, or even those who were but are no longer alive? One way to account for this is to claim that in such cases, we see and treat these objects *as if* they were alive, which minimally amounts to grasping them as having a history, as destructible, and as contributing non-instrumental value to living beings. Perhaps most important of all, the feeling of love entails that we relate to the object as vulnerable to violation, damage, and injury. From the perspective of love and recognition, this vulnerability and susceptibility to violation is, in some respects, more significant than the sheer fact of finitude or mortality. Although the latter are surely part of what it means to grasp living form, vulnerability to violation and injury are the flip side of the possibility opened up by

---

7 Kant writes: "...though respect is a feeling [*Gefühl*], it is not one *received* by influence, but one *self-wrought* [selbstgewirktes *Gefühl*] by a rational concept... Respect is actually the representation of a worth that infringes on my self-love" (GMM 4:401). The structure of respect as the mutual restriction of self-interest is key for Axel Honneth's account of the relation of recognition and how this concept develops from Kant to Fichte and Hegel. Respect "means conceiving of the value of an object in a way that compels us to set aside our mere self-interest in order to do justice to the value of that object" (2021a:104). Extending this to his Hegelian account of recognition, he writes: "the young Hegel believed that 'respect'... occurs in love almost as a matter of fact. ...forms of reciprocal recognition such as love represent modes of 'being with oneself in another'" (2021a: 125). See also 2021b: 24–25 on the importance of the idea self-limitation for recognition in Kant, Fichte, and Hegel. For a different account of the connection between love and Kantian respect for persons that focuses on the capacity for valuation, see Velleman 1999.

non-instrumental agency and interaction. If the feeling for the other as thoroughly alive opens up the possibility of a non-instrumental relationship that takes the other's self-determined ends into account, a purely instrumental relation to living form opens up the possibility not just of death — which is part of the trajectory of life — but of violation, injury, damage, abuse, and desecration. When Hegel writes that in the relation of love, living beings "against one another are in no respect dead [*von keiner Seite gegeneinander Tote sind*]," there is an acknowledgement that things can be treated as dead in different respects, such that the feeling toward the other as thoroughly alive is subject to deterioration and deformation in their ongoing interaction. Once non-instrumental interactions between living beings are opened up in loving feeling, purely instrumental interactions can threaten the thoroughly living character of the parties in the relationship. Objects of love are uniquely susceptible to violation and injury insofar are they are felt as, or treated as if, thoroughly alive.

This essential connection between love and the feeling apprehension of living form, which opens up both the possibility of non-instrumental agency and the threat that purely instrumental actions can be violent and injurious, is crucial for understanding the concept of recognition. The substantive connection between love and life extends to the relation of recognition, in which individuals reciprocally relate to others to both acknowledge and promote their self-determined ends, which in turn promotes their own. As an ethical, non-instrumental relation to another, recognition regards and treats others as thoroughly alive and in no respect dead, as purposive, dynamic, and in development, free to determine their own ends and deserving of protection from violation and injury. It is not just that love is a paradigmatic form of recognition — there can be no recognition without love as the feeling of something living, as the reciprocal feeling of living beings alike in power relating to one another as living from every point of view.

***

In light of the connection between love, life, and recognition, I want to explore two further possibilities that will contribute to the effort of expanding the concept of recognition beyond its application to the domain of human relationships: first, the possibility of proto-recognitive activities and processes in non-human life-forms; and second, the possibility of recognition and proto-recognition across life-forms, both human and non-human. What I hope to develop in rough outline is the idea of the *recognition of living form* as an ethically significant engagement with nature, broadly understood.

Can non-human life-forms engage in activities and processes that can be identified as proto-recognitive? As I suggested above, both historical and contemporary discussions of recognition generally answer this question in the negative, not primarily on empirical grounds, but on the basis of defining recognition as a reciprocal relationship between potentially equal, free, and rational persons. Although there is now overwhelming evidence supporting a wide range and degrees of cognitive and moral capacities among non-human life-forms, my aim is not to address the empirical question of whether particular kinds of creatures, or animals as a broad category, have the capacities required for recognition.[8] Instead, I will continue to develop Hegel's account of living form towards addressing the following questions: What must living form be like for relations of recognition to be possible? What is the general shape of the processes and activities of beings who are the cause and effect of themselves? What kinds of non-instrumental interactions emerge from these processes?

A being that is the cause and effect of itself must be able to produce itself in at least three ways.[9] First, it has self-organizing shape or structure in which parts and whole mutually maintain and produce one another, i.e., it exists as a living body. Second, the living body engages in processes through which it assimilates and exchanges materials with an environment to continually maintain and produce itself across its lifetime. Third, it has the power to produce another of its own kind, reproducing the species. The processes and activities at work here can be given different levels of description, for example, as mechanical or chemical processes, but there are two further kinds of activities that Hegel introduces as unique powers of living beings manifest in their processes of self-production, which tracks the forms of agency discussed above.[10] The first is "external purposiveness" or instrumental agency: in this mode of activity, the subject "exercises *violence* over the object," "takes away its specific nature

8 On the cognitive, social, and ethical capacities of animals, see Andrews 2020, De Waal 2016, and Godfrey-Smith 2020.

9 I develop Hegel's account of life in chapter seven of Ng 2020.

10 On animal agency, see Steward 2012: esp. chapter four; and Godfrey-Smith 2024: chapter three, esp. 56ff on forms of animal action. Although my discussion takes animal agency broadly as its focus, this does not entail that plant life is excluded from the picture presented here. In the *Philosophy of Nature*, Hegel discusses the history and life of the earth as a whole, calling it an "inanimate organism" (PN §341), and plant-life is expressive of a primitive form of subjectivity (PN §343). On communication and intelligence in plants and fungi see Simard 2021 and Sheldrake 2020.

[*eigentümliche Beschaffenheit*]," and "makes it into a means... [conferring] upon it its own subjectivity as its substance" (WL 6:482–483). Instrumental agency reduces the object to a mere means for the subject, treating it as dead or renders it so. The second form of agency is non-instrumental and is introduced as a possibility not for a living individual *qua* individual, but for living beings insofar as they participate in a *Gattungsprozess*, or a process of species-life. A reciprocal, non-instrumental form of activity and relationship arises as a possibility when a living being relates to another living being of its own kind. Hegel writes of this relationship that it is "immediately reciprocal [*unmittelbar gegenseitig*]" and has the character of "longing [*Verlangen*]" (WL 6:485). The relationship is immediately reciprocal because conspecifics share an identity of form. The relationship expresses reciprocal drive (*Trieb*) and longing because the relation between identical living forms generates self-feeling in relation to another, concretely expressed in species activities such as communication, competition, cooperation, mating, caring for the young, niche construction, and other collective endeavors. *Verlangen* means longing and desire but it also means request and demand, bringing us into the conceptual orbit of *Forderung* (a claim, demand, or requirement) and *Aufforderung* (a request, demand, or invitation), the latter famously employed by Fichte to develop his concept of recognition. In that context, *Aufforderung* is translated as *summons*, describing a non-coercive invitation from one free, rational being to another to exercise its free efficacy. What makes the summons and the form of relationship it initiates non-coercive is that a subject is free to act or not act in response to the invitation.[11] In this relationship, which Fichte calls "free reciprocal efficacy [*freien Wechselwirksamkeit*]," each party to the relation is engaged not as a mere means, but as a being with the power to set its own ends (FNR: 33).[12]

11 Fichte emphasizes that this relationship should not be understood in terms of causal determination: "The rational being's activity is by no means to be determined and necessitated by the summons in the way that — under the concept of causality — an effect is determined and necessitated by its cause; rather, the rational being is to determine itself in consequence of the summons" (FNR: 35).

12 For Fichte, the possibility of issuing and cognizing the summons depends upon the power of reflective judgment and the concept of an end (see FNR: 35–36). *Wechselwirksamkeit* and *Wechselwirkung* are key concepts in German idealism, and for Alexander von Humboldt, Schelling, and Hegel, the reciprocal causality of *Wechselwirkung* is closely connected with the teleological organization of life. See for example, WL 6:237–240. Werner (forthcoming) explores the concept of *Wechselwirkung* as the logical basis of the concept of recognition.

My suggestion is that Hegel's account of the non-instrumental reciprocal agency between living beings participating in the same species-process mirrors and prefigures the relationship of the summons.[13] Reciprocal longing between living beings, which here extends far beyond a relation between rational beings, is distinct from causal processes and instrumental agency for at least two reasons. First, and connected with the discussion above, reciprocal longing is possible only between beings that set their own ends. In the species activities noted above, there is not only a reciprocal feeling that conspecifics are living beings that set their own ends, but a feeling of longing that our ends may be reciprocally fulfilling: shaping my action on account of your ends may serve to realize my own. Second, and drawing on a key feature of Fichte's summons, the relation and activities of reciprocal longing are non-coercive in that they do not reduce the other to a mere means, signaled by the fact that participants remain free to *not act*.[14] Even when provoked, a competitor can back down or refuse to fight; a pursued mating partner can rebuff its pursuer; a youngling can wander away from the protective care of its parents. In the relation and agency of reciprocal longing, what is present is the free reciprocal efficacy characteristic of the summons. The activities of species-life, insofar as they express reciprocal, non-instrumental agency, are realized through *proto-recognitive relationships*. Recognition in the fully developed sense characteristic of self-conscious individuals has its basis the non-instrumental species-activities of living beings, and is possible on account of such activities. Put another way, we could say that recognition is an actualization of the relation of reciprocal longing that is a power of living beings participating in species-life, a relation in which varied forms of non-instrumental and ethical agency first become possible.

If proto-recognitive relations are manifest in the relation between conspecifics, what of the possibility of proto-recognition or recognition *across* life-forms, and especially across human and non-human forms of life? Here, although I will continue to build on ideas from Hegel and Fichte, there will also be ways in which I will depart from the strict confines of their accounts. Hegel's account of reciprocal longing opened up the structure of Fichte's summons to

13 By contrast, Khurana 2025 argues that the second-personal relations of recognition are a unique form of political sociality characteristic only of human beings, distinct from what he calls the logical and material sociality of living beings. My suggestion is that the free reciprocal efficacy characteristic of living beings participating in a species-process as such is a form of *ethical* sociality.

14 See FNR: 33.

living beings participating in the same species-process, which gives us a more precise way of understanding his claim that love is a feeling between living beings who are alike in power. Although earlier, I interpreted "alike in power" in a more open-ended manner, letting this simply refer to the power of living form as such, Hegel's distinct interest in the relation between conspecifics is evident already in another of his early writings, where he states: "love is a feeling [*Gefühl*] of a life similar to one's own, not a stronger or a weaker one" (SCF 1:363). Without denying that the relation with conspecifics enables unique possibilities of agency, or that species-life provides a necessary context for understanding and assessing dispositions, behaviors, and habits, how might we envision recognition or proto-recognition across species?

At a purely empirical level, this is in fact not so difficult to envision at all: there are abundant examples of helping behavior across species and many symbiotic relationships between species. Although the catastrophic destruction of species and ecosystems by human beings is at the forefront of much public and academic discourse, human beings are also uniquely capable of non-instrumental relationships with nature. The question is whether, conceptually, recognition can be helpful for capturing these kinds of behaviors and relationships and illuminate them in an ethically significant way. Let me begin by taking two steps toward answering this question in the affirmative. First, I think a weaker reading of being "alike in power," in which the power in question concerns living form as such, is important for understanding how recognition across life-forms is possible. If this is admitted, along with an account of reciprocity that has different modalities and degrees of activity and passivity, then the reciprocal feeling of living beings can be viewed as proto-recognitive simply insofar as it makes possible forms of agency shaped in relation to the self-determined ends of another life. It should be emphasized that it is the recognition of living form as such that makes this possible: without the recognition that the other is a subject-object who lives as the cause and effect of itself setting its own ends, there could only be causal relationships and instrumental agency, mere objects and mere means.

This brings us to a second step: up to this point, I have spoken of feeling, apprehending, grasping, and recognizing living form in a rather vague and open-ended way. But rather than needing to be resolved into a single term, it is important that the grasp of living form *is* open-ended and can be realized in many degrees and modalities, from inchoate, unconscious feelings and sensations that can distinguish between living and non-living or identify symbiotic species, to self-conscious conceptual comprehension. A self-conscious grasp of life continues to involve feeling and self-feeling, but it can also involve complex scientific theories and concepts; developed ethical relationships, obligations, and practices; and aesthetic judgment, engagement, and activity. These varied ways of grasping living form are appropriately called recognitive or proto-recognitive when the relationship engages the other *as living* — again, as a being that sets its own ends where this recognition can shape my activity with those ends in view, leaving the other free to continue in its power of self-determination. Recognition and proto-recognition of living form enables the possibility of non-instrumental, ethical relationships — both between conspecifics and across species — but it also allows us to grasp that purely instrumental relationships in such cases can be violating, injurious, and damaging. Recognition between and among living beings promotes (or at least does not impinge) the power of a living being to set its own ends, and misrecognition of life can undermine, threaten, deform, or destroy this power altogether. We can thus accept that recognition of living form has ethical significance and that recognition in an ethically significant sense across species is possible. Moreover, we can accept this while continuing to hold that there are special relationships and possibilities of agency between conspecifics, and that full-fledged forms of mutual recognition between human beings require more demanding senses of reciprocity and equality involving institutional mediation.

To conclude this brief proposal, I want to address two issues, one pertaining to love and the second pertaining to the relationship between human beings and non-human nature. The concept of love has always been central to the discourse on recognition and I began by discussing the connection between love and life proposed by the young Hegel. In conceiving of love as a feeling toward the other as thoroughly alive, I believe Hegel indeed captures something important that is part of our ordinary understanding of love, which requires that we are able to view the other as dynamic, as a being who sets its own ends, vulnerable to injury and violation. But one could also argue that this conception of love is too expansive, both undermining the distinctive qualities of love for a unique individual and requiring that a feeling of love underlies even our impersonal forms of recognition in respect and esteem. Regarding the love for a unique individual, I think the idea that the beloved is felt as alive from every point of view indeed describes what makes love so distinctive among our relationships. The connection between love and life can range from formal and general to highly specific, demanding, and concrete. On the formal end, there is Henrich's suggestion, noted earlier, that love and life are conceptually connected in exhibiting self-reference and autonomy alongside opposition and relation to otherness. There is also the general but more substantive connection between love and life in which love grasps the other as a living form that sets its own ends as the cause and effect of itself. In the concrete, specific relation to an individual beloved, the feeling for the other as thoroughly alive is a difficult, painful, surprising, joyful, enlivening, and ongoing task that is only realized among a limited number of unique others. In these relationships, seeing the other as thoroughly alive requires being able to see them anew, as having dimensions of aliveness and development that we had not previously seen, and as continually pursuing ends that shapes the lover's own purposiveness. There are of course many impersonal relationships and interactions in which we acknowledge and respect that the other sets their own ends, but do not feel or know them as living — dynamic, purposive, vulnerable — from every point of view. But what unites all of these relationships and makes them relations of recognition is that the other is treated as a possible object of love insofar as they as grasped as a living form. From the personal to the impersonal, from conspecifics to heterospecifics, being able to grasp the other as thoroughly alive, and hence, as *loveable*, is what enables relations of recognition. To avoid the impression of undue optimism here, I should also add that alongside love as a non-instrumental relation and the possibility of instrumentalizing living others, hate can also arise as a non-instrumental relation between living beings, a feeling that arises in grasping that the other sets their own ends.[15]

My main aim in this essay has been to expand and reconsider the concept of recognition to show that it can be a helpful framework for understanding ethical relationships between humans and non-human nature. I have tried to show that that recognition of living form is itself an ethically significant relation, one that allows us to relate non-instrumentally not just to other, specific life-forms, but perhaps also to natural environments, ecosystems, and even the earth system as a whole. These latter contexts and systems both contain and sustain life, contributing both instrumental and non-instrumental value to interdependent living beings that pursue self-determined ends. Recognition of living nature can thus name an ethically significant relationship in which our actions are shaped not just with the ends of other living beings in view, but in which we see the promotion of living ends as essential for realizing our own — both as individuals and as a species. Moreover, as a form of self-conscious life,

15 See Gregoratto 2025 for a critical theory of love, and especially 2025: 161–169 for an account of recognition and the activity and passivity of affect from a "critical naturalist" perspective.

human beings are uniquely capable of recognizing living form in a conceptually mediated way, capable of sophisticated understandings of the internal aims of other forms of life and can direct our behavior to, if not promote, then at least to not undermine those aims. The concept of recognition, in the expanded sense proposed here, can thus play an important role in understanding the ethical character of ecologically oriented social struggles. Expanding the concept of recognition in this way also extends the boundaries of our ethical community beyond the free association of human beings to the free association living beings on earth, an idea that is urgently needed in our present age of climate catastrophe.

References

Andrews, Kristin (2020): The Animal Mind: An Introduction to the Philosophy of Animal Cognition, Second Edition, New York: Routledge.

Bernstein, Jay (forthcoming): "From Ecocide to Nature's Right: Recognition in the Anthroprocene*," in: Matthew Congdon/Thomas Khurana (Hg.), Recognition: Historical and Philosophical Perspectives, New York: Routledge.

DeWaal, Frans (2016): Are We Smart Enough to Know How Smart Animals Are? New York: W.W. Norton.

Fichte, J.G. (2000): Foundations of Natural Right, Frederick Neuhouser (Hg.)/ Michael Baur (Übers.), New York: Cambridge University Press. (cited as FNR in text)

Godfrey-Smith, Peter (2024): Living on Earth: Forests, Corals, Consciousness, and the Making of the World, New York: Farrar, Straus and Giroux.

Godfrey-Smith, Peter (2020): Metazoa: Animal Life and the Birth of the Mind, New York: MacMillan.

Gregoratto, Federica (2025): Love Troubles: A Philosophy of Eros, New York: Columbia University Press.

Hegel, G.W.F (1991): Elements of the Philosophy of Right, Allen Wood (Hg.)/ H.B. Nisbet (Übers.), New York: Cambridge University Press. (cited as PR in text, vol. 7 of *Werke*)

Hegel, G.W.F (1975): "Love," in: T.M. Knox (Übers.), Early Theological Writings, Philadelphia: University of Pennsylvania Press, S. 302–308. (cited as FL in text, vol. 1 of *Werke*)

Hegel, G.W.F (2018): Phenomenology of Spirit, Terry Pinkard (Übers.), New York: Cambridge University Press. (cited as PhG in text, vol. 3 of *Werke*)

Hegel, G.W.F (1970): Philosophy of Nature, Part Two of the Encyclopaedia of the Philosophical Sciences (1830), A.V. Miller (Übers.), New York: Oxford University Press. (cited as PN in text, vol. 9 of *Werke*)

Hegel, G.W.F (2010): Science of Logic, George DiGiovanni (Übers.), New York: Cambridge University Press. (cited as WL in text, vols. 5–6 of *Werke*)

Hegel, G.W.F (1975): "The Spirit of Christianity and Its Fate," in: T.M. Knox (Übers.), Early Theological Writings, Philadelphia: University of Pennsylvania Press, S. 182–301. (cited as SCF in text, vol. 1 of *Werke*)

Hegel, G.W.F (1969–1970): Werke in zwanzig Bänden, Eva Moldenhauer/Karl Markus Michel (Hg.), Frankfurt am Main: Suhrkamp Verlag.

Henrich, Dieter (2003): Between Kant and Hegel: Lectures on German Idealism, David Pacini (Hg.), Cambridge: Harvard University Press.

Honneth, Axel (2021a): Recognition: A Chapter in the History of European Ideas, Joseph Ganahl (Übers.), New York: Cambridge University Press.

Honneth, Axel (2021b): "Recognition Between Power and Normativity: A Hegelian Critique of Judith Butler," in: Heikki Ikäheimo/Kristina Lepold/Titus Stahl (Hg.), Recognition and Ambivalence, New York: Columbia University Press, S. 21–30.

Ikäkeimo, Heikki (2022): Recognition and the Human Life-Form: Beyond Identity and Difference, New York: Routledge.

Kant, Immanuel (2000): Critique of the Power of Judgment, Paul Guyer (Hg.), Paul Guyer/Eric Matthews (Übers.), New York: Cambridge University Press. (cited as CJ in text)

Kant, Immanuel (2012): Groundwork of the Metaphysics of Morals, Mary Gregor/Jens Timmerman (Übers.), New York: Cambridge University Press. (cited as GMM in text)

Katz, Gal (2020): "Love is only between living beings who are equal in power": On what is alive (and what is dead) in Hegel's account of marriage," in: European Journal of Philosophy 28:1 (March), S. 93–109.

Khurana, Thomas (2025): "Living by Recognition: On the Sociality of the Human Life-Form," in: Hegel Bulletin 46.

Ng, Karen (2020): Hegel's Concept of Life: Freedom, Self-Consciousness, Logic, New York: Oxford University Press.

Ng, Karen (2021): "Humanism: A Defense," in: Philosophical Topics 49:1 (Spring), S. 145–163.

Simard, Suzanne (2021): Finding the Mother Tree: Discovering the Wisdom of the Forest, New York: Knopf.
Sheldrake, Merlin (2020): Entangled Life: How Fungi Make Our Worlds, New York: Random House.
Steward, Helen (2012): A Metaphysics for Freedom, New York: Oxford University Press.
Velleman, David J. (1999): "Love as a Moral Emotion," in: Ethics 109:2 (January), S. 338–374.
Wang, Connie (2025): "Hegel on the Concept of the Psyche, Psychic Health, and Psychic Illness," PhD Dissertation, Columbia University.
Ware, Owen (forthcoming): "Persons, Animals, and Recognition: A Classical Yoga Perspective," in: Matthew Congdon/Thomas Khurana (Hg.), Recognition: Historical and Philosophical Perspectives, New York: Routledge.
Werner, Andrew (forthcoming): "From Interacting Substances to Recognizing People: On the Logical Foundations of Hegel's Theory of Recognition," in: Matthew Congdon/Thomas Khurana (Hg.), Recognition: Historical and Philosophical Perspectives, New York: Routledge.

Algorithmic Structural Epistemic Injustice

Artificial Intelligence, Unjust Regimes of Epistemic Recognition, and the Reinforcement of Ignorance[1]

Fabian Schuppert

Introduction

Contemporary discussions suggest that we currently witness an epistemic revolution in the form of widely available artificial intelligence (AI) systems. AI systems are increasingly used in a variety of contexts, shaping the way people communicate, access information, and make decisions. This essay focuses on chatbots such as ChatGPT, which are large language models that are using a form of self-supervised machine learning (which is the artificial intelligence aspect of contemporary chatbots) in order to perform natural language processing tasks, such as natural language understanding, text classification, knowledge summaries, and natural language generation. For reasons of simplicity, for the remainder of this essay the terms artificial intelligence (AI), large language models (LLMs), natural language processing (NLP) shall all refer to particular aspects of contemporary chatbots.

Nowhere is the impact of AI-based generative chatbots more evident than in education and publishing, where these tools are deployed to write essays, recommend and summarise articles, moderate debates, grade coursework, or suggest improvements to people's writing. This means that these large language models influence which knowledge is used in certain conversations and how, as well as that they pass "judgment" on which forms of expression are "better" or "more scientific".

1 I am extremely grateful to Kalia Barkai, Hilkje Hänel, Nikolaus Hoffmann, Börries Nehe, Judith Renner, and Janina Walkenhorst for insightful comments which helped improve the argument in the paper.

According to proponents of AI-based LLMs, these technologies may seem neutral, even democratizing, as they supposedly open up knowledge to everybody, promise to remove human bias and ensure a "level playing field." However, a closer examination reveals that LLMs systematically reproduce—and even amplify—epistemic injustices and stereotypes about what counts as relevant knowledge, good writing, and rational argumentation. In addition, particular fields of inquiry are presented in very particular ways, meaning that existing stereotypes about what particular forms of knowledge can do and be, are also systematically reproduced. In doing so, AI tools perpetuate existing power structures, making the privileged even more privileged, and foster a particularly insidious form of epistemic ignorance and epistemic domination.

In their existing critiques of AI, LLMs, and algorithmic processing, scholars of epistemic injustice have primarily focused on the way that social media algorithms filter public opinion and privilege certain content over other (Stewart, Cichocki, and McLeod 2022), or the role of primarily white male programmers inscribing their own biases into the code of algorithms (Hoffmann 2019; Noble 2018), as well as how digital environments shape individual's identity formation (Origgi and Ciranna 2017). While all of these issues are serious and merit further investigation, I want to focus on a slightly different aspect, namely, the internal workings of LLMs and how their in-built regimes of recognition lead to particular forms of epistemic oppression, an issue also explored by recent contributions by Miragoli (2024) and McInerney (2024).

The Hidden Curricula of LLMs: Learning the Norms and Stereotypes of an Unjust World

Prior to being released for its intended use, all LLMs are trained on vast datasets, often scraped from published material, publicly available databases, and other forms of data that have – in one way or another – already been filtered by human hands. Instead of being a neutral pool of resources to draw from, these datasets reflect the dominant conceptions of how our world works, including problematic assumptions about what matters in the world, what typical gender roles are, and the qualities of different ethnic groups. Because of where LLMs like the chatbots this essay is concerned with are developed, there is a distinct Western bias in these models. This is hardly surprising as the LLMs are simply trained on the hegemonic norms, beliefs, and worldviews of their time, which means that also the stereotypes and blind spots of the hegemonic view are part of the training curricula. Through training, the LLMs are also supposed to be taught what counts as "valuable" knowledge, "proper" language use, "rational" argumentation, and "scientific" rhetorical style.

In other words, AI models are trained and developed on particular regimes of recognition, which bestow a particular status on some forms of knowledge, while other forms of knowledge are treated differently. What do I mean by this?

A regime of recognition is a (more or less) fixed order of who or what will get what kind of recognition for what kind of property, status or action. So, children in Germany for instance often get recognition for being "well-behaved", which is a normative marker of esteem recognition, when they say please and thank you (the actions which trigger the ascription of a form of recognition), no matter whether that really tells one anything about the kids overall behaviour. Not all forms of recognition regimes are per se morally problematic. In fact, bestowing esteem recognition and respect onto others is a central feature of human interaction (McBride 2013). But many reasons for bestowing a particular form of recognition onto someone or something are morally problematic, as in the statement "This was a really good throw for a girl". The actual esteem given initially seems positive, i.e. "a good throw", but it immediately becomes clear that the qualifier "for a girl" not only raises questions about the reason underlying the given esteem recognition, but also changes the nature of the esteem given: it is not a "good throw" simpliciter, it is a "good throw *for a* girl", which sends all kinds of messages about the status of girls when it comes to throwing. To be clear, this does not mean that this is a question of distributing recognition equally, since unequal recognition fulfils an important function in human interaction, as it is used to signal one's feelings and beliefs.

When it comes to how LLMs are trained, it is by no means the standard case that some forms of knowledge get a lot of recognition, while others get very little. While this might be a case of problematically recognising different forms of knowledge, it is much more common that different forms of knowledge also get different kinds of recognition, which reflect and perpetuate problematic stereotypes about what specific forms of knowledge are and can be.

It is therefore not the case that AI generated essays on the status of "indigenous knowledge" would be either using overtly negative language or directly express the view that indigenous forms of knowledge are worth less than other forms of knowledge.[2] Instead statements on indigenous knowledge regularly included adjectives such as "spiritual", "traditional", and "cultural", while essays on feminism include terms such as "care", "subjective", and "emotion".

Needless to say, all of the above terms could form part of excellent treatments of indigenous knowledge or feminism, since indigenous scholars do indeed often highlight the role of traditions, cultural values, and spiritual worldviews, while feminists have long advocated for the importance of subjective experiences of oppression, the centrality of care work, and the role of emotions in reasoning. The problem with the examined essays is that these terms are used in juxtaposition to terms like "rational", "scientific", "established" and with insufficient critical contextualisation, so that implicitly indigenous knowledge apparently never can be "rational". Put simply, indigenous knowledge is put into a box with a very small number of labels attached, labels which reproduce hierarchies of esteem, as long as being rational, scientific and established continue to be seen as the ideal norm that all other forms of knowledge and research aspire to. The issue is thus not that indigenous knowledge is not esteemed or recognised in LLM-generated content, but that it is esteemed and recognised only in very particular ways and that it is portrayed in reference to what counts as "normal proper knowledge".

Interestingly a similar pattern can be observed when LLMs are tasked with "improving the writing" of a particular text or assessing the "rationality of an argument". "Good writing," as assessed by LLMs, is often synonymous with the conventions of standardized English and the pseudo-academic prose found in executive summaries of scientific reports, which try to sell a particular solution. Hence, there is a particular tone of expression, somewhere between marketing pitch and policy report summary that is seen as the gold standard of "good writing". Norms of rational argumentation are typically equated with forms of argumentation that privilege deductive reasoning and that use marker terms such as "objective" and "evidence-based". Expressions such as "subjective experience" and "informal norms" were taken out of pre-written inputs and replaced by LLMs by expressions such as "empirical evidence" and "well-known facts", which clearly express a different thing than the initial expressions.

---

2 The following empirical observations are based on a three-month trial using different prompts for three widely used chatbots: ChatGPT, Google Gemini, and Microsoft Co-Pilot.

When it comes to hidden training curricula of standard LLMs, two things happen: first, the algorithms are trained on hegemonic norms and worldviews, which come with an in-built regime of recognition that distributes esteem markers to all sorts of agents, views, events, and states of affairs. While a critic might point out that implicitly passing on markers of esteem is a necessary aspect of any curriculum, it makes a huge difference on which data and knowledge a model was trained, and which recognition order has been passed on. As is well-established, predominant views of history, politics, and society are often racist and sexist, which means that algorithms trained on this data suffer from the same problem. LLMs do not merely reflect but encode and reinforce prevailing norms—norms that are themselves the product of longstanding histories of inclusion, marginalisation, exclusion, and epistemic silencing.

Taking privileged norms and worldviews as the "neutral" knowledge base for the most widely used chatbots and LLMs leads to a perpetuation of privilege. There is a distinct epistemic injustice already at the training and input level of most LLMs, since the in-built recognition regime that structures what counts as what expresses a whole range of identity stereotypes and problematic value judgments which affects the normative status of different kinds of knowledge and agents as knowers.

The internal workings of LLMs: Recognition orders, social capital, and epistemic authority without sufficient reason

The epistemic injustice at the training stage is not the only input injustice of most LLMs. The second input injustice occurs through how LLMs like ChatGPT actually work. While many people believe that the latest generation of LLMs blaze through the entire vast data that can be found on the internet in order to generate an answer to a query within seconds, the truth is actually quite different. Similar to how search engines like google operate, ChatGPT, Gemini and co. use shortcuts, by operating with directories of webpages, which through web-crawling have been deemed to be statistically particularly relevant. In other words, when one sends a short prompt to a chatbot (e.g. "What is climate change?") the program does not search the entire web and then passes judgment on which information to provide as an answer; the search in the web would not only take much too long, but also the program is not able to pass an evaluative judgment that would be sensitive to all sort of different queries.

The algorithms at work in contemporary LLMs of course do pass value judgments in some sense, since what the algorithms are trained to do is to draw their material from the pages/sources of information which the underlying metric (which was already used as part of the web crawling) to be statistically the most relevant. All that matters during the process of web crawling and during the actual query to the LLM is that the information on a website has been "easily discoverable" through web search optimization and that the site is "well networked", that is it is connected by a myriad of links from other webpages.

There are two distinct problems here with the internal workings of LLMs: first, the *regime of recognition* that underpins the determination of which sources of information are relevant, is based on a form of *privilege and social capital*; second, *the ascription of epistemic authority* that AI-based LLMs happens *without an appropriate set of good reasons*.

Let me start with the first problem, the regime of recognition which is based on a form of privilege and social capital. As mentioned above, LLMs rely on a register of webpages that were discovered through web crawl and that were deemed to be statistically particularly relevant. This means that these webpages get a lot of hits during the web crawl process, in part because these webpages were optimised for this kind of crawl (e.g. search engine optimisation – SEO). It is the same system that advanced search engines like google rely on. What this means is that you can game the system and make sure that your webpages get a lot of hits. One particularly easy way to do this is by spending money on it, which is unsurprisingly something that privileged actors can do much more easily than other actors. It also helps if you have a large portfolio of sites which constantly refer to each other, but which do not look like they come from the same tree. Springer Academic Publishing is really good in this area, which means that articles published in their journals, or chapters published in their books get a lot of hits on google and they also feature heavily when chatbots are tasked with writing a scientific essay on a particular topic. The actual quality of the article or chapter is irrelevant, since the LLMs' esteem recognition for a particular source of knowledge is based on its statistical hit-rate.

This is where the second part of the first problem comes in, namely the importance of "social" networks, that is, many external sites which refer to the page one wants to optimise for discovery. For external sites to refer to one's own page, though, one needs to be well connected. Just like with human beings then, social networks matter, which means this case could be interpreted as a case related to what Bourdieu called social capital, namely, the benefits individuals and groups derive from social connections, networks, and relationships (Bourdieu 1986).

In order to be statistically relevant, then, it helps if one is economically privileged, as well as privileged regarding the social capital one has at one's disposal. Because of how LLMs are designed, they rate these privileges highly and reward well-connected and search optimised sites with more esteem recognition, which makes it more likely that any future query will be answered by a chatbot using the information provided by privileged agents.

This reproduction of privilege is in many ways subtle but extremely profound. Precisely because LLMs are built to award esteem recognition to those pages that get a lot of hits in their metrics, LLMs are extremely liable to be tricked into assuming that the sites which are best at being highlighted by a web crawl and a chatbot prompt, are also the best sources of knowledge. This is precisely the second problem, i.e. the ascribing of epistemic authority without sufficient reason.

When processing a prompt, chatbots do not care about the academic credentials of the author of the webpage the chatbot draws the answer from. This can be on the one hand refreshing, since not everything coming from an Oxbridge educated person gets automatically treated differently than everyone else's views. But on the other hand, since LLMs *only care* about statistical relevance there is no quality control outside of statistical relevance. There is no quality control mechanism in the sense of an algorithm passing evaluative judgment on what a most complete and sophisticated answer should look like. Instead, whatever is statistically most relevant is taken to be objectively good information, simply because it is statistically most relevant. Statistical relevance on its own, though, is not a good enough reason for ascribing something epistemic authority, especially in light of the reproduction of privilege and social capital, which I described above.

The Epistemic Injustices of LLM Outputs

For a long time, the importance of statistical relevance within the internal workings of LLMs was directly reflected in the output of Generative Pre-Trained Transformers (GPTs). As Emily Bender and colleagues (2021) provocatively asked as recently as 2021, will even the most sophisticated GPTs ever be more than "stochastic parrots" that – while being able to process natural language in such a way that one can have a conversation with them – will only ever repeat information back to their human interlocutors that they were initially trained with or that they copied and pasted from a webpage?

As Konstantine Arkoudas (2023) argues, Bender's question has been answered with a resounding "yes", since ChatGPT and the latest generation of chatbots are much more than stochastic parrots. The latest GPTs can come up with their own views and arguments, combining different sources of information and developing (relatively) independent conclusions, which goes even as far as inventing concepts or pieces of literature that the GPT has inferred should exist. As Arkoudas (2023) points out, this is a huge technological advancement, but it does not mean that the latest GPTs are indeed "intelligent reasoners", since the latest chatbots still struggle with exercises in logic, text-based maths problems, and the identification and application of norms. At their core, the latest chatbots are still LLMs, which try to identify the correct answer by relying on statistical relevance and the regimes of recognition described above.

Considering the different epistemic injustices that we could identify at the input stage of LLMs, it is hardly surprising that we can also observe epistemic injustices at the output stage. In the following, I want to focus on three issues, all of which are directly related to the perpetuation of privilege and ignorance discussed above: epistemic oppression, silencing, and toxic deficiency.

Epistemic oppression: Building on Kristie Dotson's (2014) account of epistemic oppression, one can identify how the recognition order underlying statistical relevance which is tied up with privilege, social capital, and the unjust background of a long history of biased framing and naming, leads to a situation in which marginalised groups and forms of knowledge are systematically oppressed. Because of how LLMs are set up and work, with their unjust input and their focus on producing answers that reflect statistical relevance, the output of the latest chatbots fails to adequately reflect and incorporate the knowledge and insights of many marginalised groups. For example, it is not the case that on the internet there are no sources on Black feminist thought, but when one looks at the results of chatbot queries regarding important 20th century social thought, or even feminist thought, these sources – most often provided by Black female writers – are strangely absent. This means that in the realm of LLMs the contribution to knowledge by Black feminist writers is made virtually impossible.

Toxic deficiency: Toxic deficiency is a phenomenon aptly described by Martin Miragoli (2024, p. 9) who points out marginalised groups are not just harmed as knowers (as in the case of epistemic oppression) but also as knowledge seekers. Because of the epistemic injustices occurring at the input stage of LLMs, LLMs have in-built "hermeneutical lacunae" (Miragoli 2024, p. 10), in that within the shared hermeneutical resources of the LLM universe concepts and understandings pertinent to the lived experiences of marginalised groups are missing. This leads to a toxic deficiency in the outputs of LLMs, in which the hermeneutical lacunae are translated into "answers" which are deficient conceptually and which try to reinforce hegemonic norms onto the person who seeks knowledge. A young trans-person trying to make sense of their identity and feelings will often encounter said deficiency, while at the same time being confronted with heteronormative binary gender stereotypes, which can aptly be described as toxic, since they hurt the young person's sense of self. The same is true for a Black person, who wants to know more about the history of Black civilisation which in mainstream history-writing has been rendered invisible. This invisibility is on the one hand a deficiency and on the other hand it is covered by a toxic array of presentations of history in which entire parts of the world are treated as blank spaces which only come into view when European imperialism demands it. Black civilisation is not only conspicuously absent, it is negated through toxic colonial imaginaries.

Silencing: A third issue is silencing (Dotson 2011), which happens mainly when we turn our attention to the linguistic preferences of contemporary chatbots, which do not give all speakers the same recognition, because some expressions and ways of conversing are deemed to be less desirable and sophisticated. By privileging dominant norms of writing, speaking, and rationality, AI systems help maintain the social and epistemic dominance of those who already wield it. Students, professionals, and creators who already write and argue in the expected style are further rewarded, while those who do not are penalized and silenced. The privileged thus become even more privileged, as their modes of expression are held up as universal standards.

With the spread of AI-supported LLMs into various areas, this problem becomes even more pronounced: automated grading systems in schools can systematically disadvantage students who speak in dialects or bring nontraditional modes of argumentation to their work. In publishing, algorithms that recommend or surface content based on "readability" or "argument quality" quietly filter out diverse voices.

Following the description of these three kinds of injustice, two very important points need to be made: first, the observed injustices are not accidents or a malfunctioning of otherwise well-working systems, but they are a systematic design-feature (Ruiz and Sertler 2024; Miragoli 2024); second, the kinds of epistemic injustices observed are not based on personal interactions, but forms of systemic epistemic injustices.

Let me start by stressing that all three injustices at the output stage are systemic issues, as Miragoli (2024) convincingly argues regarding toxic deficiencies:

> "That is: because the hermeneutical lacuna present in our shared online resources is a direct consequence of the very functioning and training practices of ML-based AIs, epistemic injustices of a hermeneutical kind [...], are not just unlucky byproducts of developers' biases, but a *systematic feature of the design of AI design*." (Miragoli 2024, p. 11)

The same is true when it comes to epistemic oppression, since certain groups and sources of knowledge are systematically excluded at both input stages, as well as at the output stage, making it thus impossible for some knowers to contribute to the shared wisdom of society. Contemporary chatbots/LLMs thus produce epistemic injustices *by design*, and not just as an unintended side product.

Secondly, the kinds of epistemic injustices observed are not based on personal interactions, but forms of systemic epistemic injustices, which is what distinguishes the cases described here from most of the cases described by Dotson (2011; 2014). Unlike in interpersonal cases, where for instance testimonial injustices are based on identity prejudices against a particular speaker who is harmed in their capacity as a knower, in the case of silencing LLMs simply process an available text trying to improve its writing or grade student essays, which cannot be clearly attributed to a particular group or speaker; LLMs thus commit an epistemic injustice in a slightly different way: LLMs come with in-built recognition hierarchies which assign epistemic status and epistemic validity based on token markers and statistical relevance, which are supposed to generally signal what should count a sound argument or as a proficient use of language. The same recognition regimes in combination with metrics of statistical relevance are used by LLMs to determine what an objective and truthful answer should be.

The Cultivation of Epistemic Ignorance

Ironically, in their quest to produce "objective" or "truthful" answers, LLMs actually generate the opposite; these tools cultivate a particular kind of epistemic ignorance—one that is both pervasive and difficult to detect, because it follows in the footsteps of hegemonic views on knowledge, norms, and history, which have long claimed to simply advance the truth and nothing but the truth.[3]

One particularly pernicious effect of widespread AI-based LLM use in contemporary chatbots is the homogenisation of knowledge. LLMs prune away stories and artefacts that sit uncomfortably with the established consensus. Therefore, chatbots prompted to write short five-page essays about African and European history, paint very different pictures of what happened in the 20th century. The essay on African history focused on civil strife, military coups and economic challenges in 20th century Africa, while Europe was presented as the beacon of human rights and solidarity, with a special positive mention of the European Union. This is of course how much of mainstream media represents both continents, so why should this be a particular concern for scholars of epistemic injustice? The reason is that we can observe here how AI tools perpetuate epistemic injustices and forms of misrecognition systematically, despite these tools having been trained to avoid identity prejudices based on superficial markers. Developers of LLMs forcefully argue that their algorithms avoid identity-based biases and use "neutral" metrics for assessing the "value" of a piece of information, that is, statistical relevance and a wide network. As shown above, however, these are not sufficiently good reasons to ascribe something general epistemic authority.

LLMs thus promise truth and objectivity, but what they generate is a false or weak objectivity (Harding 1995). Users may believe that algorithmic recommendations and assessments are neutral, when in fact they are deeply implicated in longstanding hierarchies of race, class, language, and knowledge, precisely because the markers of statistically relevant value and esteem are in themselves coded in racist, sexist, and classist ways.

This reproduction of privilege is in many ways subtle but extremely profound. It ensures that those at the centre of cultural, linguistic, and epistemic power have their perspectives normalized and reaffirmed, while those at the

3 Much of the "knowledge"-base of LLMs is itself guilty of what Gaile Pohlhaus Jr. (2012) has called wilful hermeneutical ignorance.

margins are forced to translate, assimilate, or remain unheard, while still being oppressed at the same time. This problem is particularly pronounced in cases where long-standing practices of silencing, cultural exclusion and historic whitewashing exist and have led to a state, in which alternative accounts of history have been systematically oppressed.

Because latest generation chatbots operate under the assumption that there always is one correct and objective answer, and because the algorithms equivocate statistical relevance with epistemic authority, contribute to a homogenization of knowledge. In this regard, Google and ChatGPT are very much alike, since they try to find "canonical" results, assuming that reducing complexity is in this case a good thing. This is precisely where LLMs ought to learn from Harding's (1995) nuanced account of strong objectivity, which aims to block "might is right" accounts of knowledge. Maximising objectivity requires a wide and diverse knowledge base, which is where the regimes of recognition of existing LLMs fall down, because they breed ignorance through advancing a narrow understanding of objectivity.

Reducing Epistemic Injustice through training LLMs?

At this point, critics might object that my view of LLMs is too negative, since ChatGPT and friends do bring a range of benefits: people do use chatbots for improving their English/their writing, and it is possible to train latest generation chatbots through prompting them repeatedly. While it is of course true that chatbots can help people in all sorts of ways, this does not change my primary point, namely, that LLMs by design bring a range of systemic epistemic injustices with them.

In addition, the idea that conversing with chatbots is a good way to train them to avoid future injustices is misguided and a bit naïve. The reason for this rather harsh assessment is again the way that chatbots actually work. Chatbots are indeed designed to be further trained through conversations with human end-users. It's what is called Reinforcement Learning from Human Feedback (Christiano et al., 2017), which means that chatbots adapt to what their interlocutor is searching for. Therefore, it is entirely possible that a Black feminist might make their chatbot sound more like a Black feminist chatbot, by asking all sorts of challenging questions and forcing the LLM running the chatbot to look for particular information. But what happens here has nothing to do with reducing epistemic injustice.

First of all, training a chatbot is only possible if one already possesses the relevant knowledge of what one is looking for. In other words, one needs to be aware of the deficiencies, in order to be able to train.

Second, in training the chatbot, one actually provides free labour and one offers one's epistemic capacities freely for capitalist extraction. Basically, the Black feminist would have turned themselves into a free resource.

Third, chatbots do not adjust their general answers, but only in this particular conversation, which means what we can observe here is more a case of mirroring or disingenuous talking back, which basically would function like a social media echo chamber.

What remains untouched is the underlying structure, in which hegemonic statistical dominance is what matters. This means that what really works best for changing chatbot outputs, is to "flood the zone with shit", which is a problem one can already observe. There have been repeatedly cases in which AI-supported LLMs have been tricked by AI-generated content, leading to a self-reinforcing circle of fake claims, masquerading as objective fact. As Microsoft's disaster with their chatbot Tay showed, feeding a chatbot huge quantities of certain forms of information can make a difference, but this works best if the information provided is extremely simple, which runs counter to what a feeding of information in the name of epistemic justice would require.

Conclusion

My aim in this short essay was to highlight a few of the distinct epistemic injustices that AI-supported LLMs raise. While contemporary LLMs are trained so as to avoid clearly racist, homophobic, and sexist language, the regimes of recognition with which LLMs operate at the input stage lead to the reproduction of privilege and ignorance. As a result of this unjust background, the output of LLMs often leads to forms of epistemic oppression, toxicity, and silencing. However, these pernicious effects are not unfortunate accidents or a malfunctioning of the highly sophisticated AI-supported LLMs, but they are systematic features by design (Miragoli 2024; Ruiz and Sertler 2024). Therefore, hopes that end users can train LLMs in such a way as to produce epistemically just chatbots are overly optimistic and naïve. While making LLMs' internal workings more transparent is certainly a step in the right direction, it is the systemic design and its regimes of recognition that really is the root problem.

References

Arkoudas, Konstantine (2023). "ChatGPT is no stochastic parrot. But it also claims that 1 is greater than 1." *Philosophy & Technology* 36.3: 54.

Bender, Emily M., et al (2021). "On the dangers of stochastic parrots: Can language models be too big?." *Proceedings of the 2021 ACM conference on fairness, accountability, and transparency*. March. pp. 610–23. https://doi.org/10.1145/3442188.3445922.

Bourdieu, Pierre (1986). "The forms of capital". In: Richardson J (ed.) *Handbook of Theory and Research for the Sociology of Education*. New York: Greenwood, pp. 241–58.

Dotson, Kristie (2011). "Tracking epistemic violence, tracking practices of silencing." *Hypatia* 26.2: 236–257.

Dotson, Kristie (2014). "Conceptualizing epistemic oppression." *Social epistemology* 28.2: 115–138.

Harding, Sandra (1995). "'Strong objectivity': A response to the new objectivity question." *Synthese* 104.3: 331–349.

Hoffmann, Anna Lauren (2019). "Where fairness fails: data, algorithms, and the limits of antidiscrimination discourse." *Information, Communication & Society* 22.7: 900–915.

McBride, Cillian (2013). *Recognition*. Polity.

Mclnerney, Tomás (2024). "The Algorithmic Construction of Epistemic Injustice." *Socio-Legal Studies of Epistemic Injustice and Spaces and Places*.

Miragoli, Martin (2024). "Conformism, Ignorance & Injustice: AI as a Tool of Epistemic Oppression." *Episteme*: 1–19.

Noble, Safiya Umoja. *Algorithms of oppression: How search engines reinforce racism*. New York University Press, 2018.

Origgi, Gloria, and Serena Ciranna (2017). "Epistemic injustice: The case of digital environments." *The Routledge handbook of epistemic injustice*. Routledge, 303–312.

Pohlhaus Jr, Gaile (2012). "Relational knowing and epistemic injustice: Toward a theory of willful hermeneutical ignorance." *Hypatia* 27.4: 715–735.

Ruíz, Elena, and Ezgi Sertler (2024). "Injustice by design." *Inquiry*: 1–24.

Stewart, Heather, Emily Cichocki, and Carolyn McLeod (2022). "A perfect storm for epistemic injustice: Algorithmic targeting and sorting on social media." *Feminist Philosophy Quarterly* 8.3/4.

Contributors

Clara Berlich researches in aesthetics, feminist and political philosophy. Her doctoral project explores the phenomenon and concept of resistance through an analysis of the reception history and various artistic interpretations of Sophocles' *Antigone*.

Jacob Blumenfeld is a postdoctoral fellow at the Centre for Social Critique at Humboldt University, Berlin, where he co-directs an interdisciplinary research project on socialization. He is the author of *The Concept of Property in Kant, Fichte, and Hegel: Freedom, Right, and Recognition* (Routledge, 2023) and *All Things Are Nothing to Me: The Unique Philosophy of Max Stirner* (Zero Books, 2018). He is the co-editor (with Niklas Angebauer and Tilo Wesche) of *Umkämpftes Eigentum: Eine gesellschaftstheoretische Debatte* (Suhrkamp, 2025) and (with Chiara Bottici and Simon Critchley) of *The Anarchist Turn* (Pluto Press, 2013).

Hilkje C. Hänel teaches and researches at the University of Potsdam in political and social epistemology, feminist philosophy, political theory and applied ethics with a focus on migration and disability. She is the author of several papers and books, including a recent book on epistemic injustices and epistemic oppression, and co-editor of two book series as well as the *Journal of Social Ontology*.

José Medina is Walter Dill Scott Professor of Philosophy at Northwestern University. His primary fields of expertise are critical philosophy of race, feminist and queer theory, social epistemology, and political philosophy. Medina's latest book is *The Epistemology of Protest: Silencing, Epistemic Activism, and the Communicative Life of Resistance* (OUP, 2023). Previous books include *The Epistemology of Resistance: Gender and Racial Oppression, Epistemic Injustice, and Resistant Imagina-*

tions (OUP, 2013), recipient of the North-American Society for Social Philosophy Book Award.

Karen Ng is Associate Professor of Philosophy at Vanderbilt University. Her research focuses on post-Kantian philosophy and critical theory. Her first book, *Hegel's Concept of Life: Freedom, Self-Consciousness, Logic*, was published by Oxford University Press in 2020 and won the 2021 *Journal of the History of Philosophy* Book Prize.

Breno Santos is philosophy professor at the Federal University of Mato Grosso (UFMT) in Brazil, mainly working on Social and Political Epistemology, Philosophy of Science, and Political Theory. He's the Head of the Philosophy Graduate Program at UFMT and General Secretary to the local branch of the Teachers' Union.

Fabian Schuppert is Professor of Political Theory at the University of Potsdam, Germany. Fabian works on issues in political philosophy, climate ethics, as well as political economy.

Peimaneh Yaghoobifarah is a teacher based in Berlin. Her academic work explores the intersection of epistemic injustice and theories of recognition. Her practical experience as a teacher and political educator in schools grounds her analysis of how children, especially students, face epistemic injustice in the German school system.